T0185634

# Food Microbiology and Food Safety

## Practical Approaches

**Series Editor:**

Michael P. Doyle
Regents Professor of Food Microbiology (Retired)
Center for Food Safety
University of Georgia
Griffin, GA, USA

# Food Microbiology and Food Safety Series

The Food Microbiology and Food Safety series is published in conjunction with the International Association for Food Protection, a non-profit association for food safety professionals. Dedicated to the life-long educational needs of its Members, IAFP provides an information network through its two scientific journals (Food Protection Trends and Journal of Food Protection), its educational Annual Meeting, international meetings and symposia, and interaction between food safety professionals.

## Series Editor

Michael P. Doyle, *Regents Professor of Food Microbiology (Retired), Center for Food Safety, University of Georgia, Griffin, GA, USA*

## Editorial Board

Francis F. Busta, *Director, National Center for Food Protection and Defense, University of Minnesota, Minneapolis, MN, USA*
Patricia Desmarchelier, *Food Safety Consultant, Brisbane, Australia*
Jeffrey Farber, *Department of Food Science, University of Guelph, ON, Canada*
Vijay Juneja, *Supervisory Lead Scientist, USDA-ARS, Philadelphia, PA, USA*
Manpreet Singh, *Department of Food Sciences, Purdue University, West Lafayette, IN, USA*
Ruth Petran, *Vice President of Food Safety and Pubic Health, Ecolab, Eagan, MN, USA*
Elliot Ryser, *Department of Food Science and Human Nutrition, Michigan State University, East Lansing, MI, USA*

More information about this series at http://www.springer.com/series/7131

Hal King

# Food Safety Management Systems

Achieving Active Managerial Control
of Foodborne Illness Risk Factors in a Retail
Food Service Business

 Springer

Hal King
Saint Simons Island, GA, USA

ISSN 2626-7578 ISSN 2626-7586 (electronic)
Practical Approaches
ISBN 978-3-030-44737-3 ISBN 978-3-030-44735-9 (eBook)
https://doi.org/10.1007/978-3-030-44735-9

This Springer imprint is published by the registered company Springer Nature Switzerland AG
The registered company address is: Gewerbestrasse 11, 6330 Cham, Switzerland

*This book is dedicated to my mother, Bobbie King Patrick, who made sure I went to college (I wasn't planning on it nor ready for it), gave me time to figure it out, was always there cheering me on, and to this day is my biggest fan. I remember when I was in middle school after my father passed away, the farthest thing on my mind was improving my reading and comprehension abilities; all I wanted to do was play my drums and play outside. However, my mom took me to reading comprehension classes every week to help me improve my reading comprehension skills and supported all of my interest in science then and now. I did not know it nor appreciate it at the time, but it was the very thing that enabled me to learn and think in graduate school enabling a "drummer" to achieve a Ph.D. in Medical Microbiology and Infectious Diseases. This one act of love initiated a career in public health that has helped me be a life-long learner – as I now attempt to use my experience and learnings to help others.*
*Thank you, Mom!*

# Foreword

What's at stake when restaurant operators don't take food safety seriously or as you may have heard your own staff or even your leadership ask, "Why we gatta do this?" Certainly, the statistics bear repeating: Foodborne illness costs the United States about $152 billion/year, each year approximately 1 in 6 Americans get foodborne illness, foodborne illnesses result in over 3000 deaths each year, and on average year to year 60% of foodborne illness outbreaks occur in restaurants. While statistics *are* important, real-life examples are always compelling. We might say that the modern era of restaurant food safety began with the Jack in the Box *E. coli* O157:H7 outbreak in 1993. Although there had been 22 documented outbreaks of *E. coli* in the United States prior to the Jack in the Box incident, it was the high-profile Jack in the Box event that put the need for restaurant food safety management on the map. Seven hundred and thirty-two people were infected, 4 children died, and 178 other victims were left with permanent injury including kidney and brain damage.

Fast forward to 2013 where Chi Chi's filed chapter-11 bankruptcy following a Hepatitis A outbreak; where 4 people died, 650 were sickened, and 9000 had to be immunized. Likewise, Chipotle, a company that built its reputation on selling food with integrity, ran a gauntlet of unfortunate foodborne illness events for several months in 2015–2016. Many were sickened, and company stock lost nearly half its value. Happily Chipotle has rebounded thanks in part to the implementation of solid food safety leadership and management. There are also additional costs to a restaurant's business in employee morale, training replacements, or retraining existing staff. Litigation cost Jack in the Box over $98 million dollars. The impact on a brand's reputation is magnified in today's connected world of social media. In such an atmosphere we can certainly understand what Warren Buffet meant when he suggested that it takes 20 years to build a reputation and 5 min to ruin it. Thoughtful restaurant operators today understand why it's important to control foodborne illness risk, and can answer when asked, "Why we gatta do this?"

According to Annex 4 of the FDA Food Code, "Active Managerial Control (AMC) means the purposeful incorporation of specific actions or procedures by industry management into the operation of their business to attain control over foodborne illness risk factors. It embodies a preventative rather than reactive approach to food safety through a continuous system of monitoring and verification." Those foodborne illness risk factors, identified by the Centers for Disease Control and Prevention (CDC) as being responsible for the greatest number of foodborne illnesses and outbreaks, are poor employee health and hygiene, improper holding temperatures, contaminated utensils and equipment, improper cooking temperatures, and purchasing from unsafe sources. Food Safety Management Systems (FSMS) are necessary to control these risk factors. There are several elements of all FSMS designed to achieve Active Managerial Control over these risk factors. They are written food safety policies and procedures, training on those procedures, monitoring, corrective actions, management oversight, and the periodic re-evaluation of the process. One of the most effective ways to achieve Active Managerial Control of the foodborne illness risk factors using FSMS in a retail foodservice setting is through a method known as Process HACCP. In contrast to food manufacturing where only one product at a time is produced on an assembly line, restaurants produce multiple food products in the same prep area, and often at the same time. By using Process HACCP, the many variables found in the preparation of food in restaurants can be divided into food preparation processes based on the number of times a given food's internal temperature passes through the Temperature Danger Zone (TDZ) of 41–135°F.

There are few people better qualified to help us digest this soup of acronyms and navigate the interconnection between FSMS and Active Managerial Control than my friend and colleague, Dr. Hal King. Among food safety professionals, Hal is the point of the spear.

Hal's approach in his new book is not theoretical, as he has led food safety management for a large foodservice business and implemented effective FSMS in its restaurants. This book includes ideas for design and implementation of FSMS in your foodservice establishment or within those of your corporate business. Prerequisite programs are not overlooked, and neither are the roles of training and facility design.

The first thing to acknowledge about any new business model, such as implementing FSMS in a restaurant, is the need to establish it in the business's culture in order to sustain it; and culture is complex. It's often assumed that culture in an organization is homogenous and therefore more or less effective, but organizations are made up of subgroups; in other words, cultures have subcultures. Think of a baseball team: infield vs outfield, middle infield vs corner infield, pitchers and catchers but then starting pitchers, middle relief pitchers, and closers. In such a landscape, a set of shared values, attitudes, and practices doesn't fall into place overnight. It involves many elements and requires practice, execution, repetition, and perseverance. I refer you to a quotation by Hal's mentor, the late S. Truett Cathy, cited in this book. "Repetition yields constants, constants create cultures." Creating a culture of consistent execution is important for delivering your brand promise, ensuring a positive customer experience, *and* for controlling foodborne illness risk. Food Safety Management Systems used daily to achieve Active Managerial Control will help to ensure that consistency and culture.

Building relationships between the restaurant industry and the regulatory authorities is also mission-critical if we hope to effectively work together to reduce the risks of foodborne illness. We must accept one another as partners with a common purpose: to protect the health of the dining public. This benefits all stakeholders including the consumer, protects your brand, and helps control the risk of foodborne illnesses and outbreaks. Most of the important knowledge to prevent foodborne illnesses in foodservice comes from government agencies like the FDA and the CDC. One of the ways we *can* work together and learn to view one another as partners is by building relationships among stakeholder groups in *safe-harbor* environments where benefits of collaboration but also barriers to collaboration and overcoming those barriers can be candidly discussed. Where these relationships exist, greater collaboration towards continuously improving the safety of the food supply and public health becomes possible. Where these relationships are absent, the food industry is left to *react*, which can breed a culture of fear, *especially* during inspections. This in turn creates a combative atmosphere between inspector and restaurant operator which actually inhibits the creation of environments conducive to learning, and to the control of risk factors. Quoting from a Fairfax County, Virginia, survey of restaurant operators who were using a local health department sponsored Active Managerial Control program, "Before, it was more like a cop waiting in a speed trap to bust you. Now, it is more like a partnership between us and the inspector, with the common goal of providing a healthier and safer food environment." As you can read in the Appendix of this book, Fairfax County, Virginia, provides one example of a collaborative partnership that brings industry and regulatory stakeholders together instead of wedging them apart. Forward thinking program managers in jurisdictions like Fairfax County are enabling industry to succeed in its implementation of Active Managerial Control, and they should be congratulated and emulated. But the food industry in other jurisdictions across the United States must not wait. Become ambassadors for your brand and share your Food Safety Management System for controlling foodborne illness risk with your regulatory authority.

As food safety professionals we don't practice our professions in a vacuum. We all understand the need for buy-in from business owner/leadership or the C-suite. In his final chapter, Hal articulates a value proposition for using Food Safety Management Systems to achieve Active Managerial Control that should resonate with a CEO, COO, or CFO. Hal looks at each component part of FSMS as a way to build the value-case for a business. Preventing foodborne illness, limiting the associated costs of foodborne illness mitigation, preventing health inspection violations and increasing inspection scores, addressing false claims and legal actions, and enhancing business outcomes can all be monetized. Our ability to communicate value to senior executives can make or break any attempt at implementation of an effective, robust FSMS, and Hal's chapter is one of the most compelling arguments I've read. Food safety professionals and/or business leadership that understand how to have these conversations are well on their way to establishing a robust Food Safety Management Program and the necessary FSMS in their restaurants while those that don't will continue to gamble with the risk of causing foodborne illnesses and outbreaks. I view it as indicative of Hal's passion for your business success that he would devote a chapter to this topic.

This book is erudite throughout and loaded with extensive references. Yet, for all its scholarship, it is an accessible read that can be used as a tool by anyone who is looking to build robust FSMS or validate one already in place.

If indeed a rising tide lifts all boats, Dr. Hal King's new book is on the flood tide.

Food Safety & Quality Assurance Programming,                          Mark S. Miklos
National Restaurant Association
Washington, DC, USA

# Preface

Foodborne disease outbreaks caused by foods prepared and sold in foodservice establishments (i.e., restaurants and any other retail foodservice businesses that prepare food for immediate consumption, and sale it directly to a consumer) continue to be a major public health issue in the United States. The Centers for Disease Control and Prevention (CDC) states (CDC 2019) that over 64% of the foodborne disease outbreaks that have occurred in the year 2017 in the United States (on average it is 60% every year; 2017 is the most current CDC report that analyzes location of foodborne illness) were linked to restaurants. There were **489** foodborne disease outbreaks caused by restaurants in 2017 with the largest percentage, interestingly, coming from sit-down dining restaurants not fast-food restaurants (Table 1).

Looking at all foodservice establishments (not just restaurants but including foodservice establishments in colleges/universities/schools, hotel/motels, catering businesses, banquet facilities, festivals/fairs/food trucks), there were **653** outbreaks, **10,696** illnesses, **343** hospitalizations, and **5** deaths in 2017 (the latest data available from the CDC National Outbreak Reporting System (NORS) (https://wwwn.cdc.gov/norsdashboard/, 2020). There are numerous more foodborne disease outbreaks and sporadic cases of illness than reported to the CDC as many occur within states that do not report small outbreaks, are sporadic cases from a foodservice establishment but not associated with an outbreak (see **Chap. 2**), occur in foodservice loca-

**Table 1** Foodborne disease outbreaks and outbreak-associated illnesses by location of food preparation from the Foodborne Disease Outbreak Surveillance System, USA, 2017

|  | Outbreaks | |
| --- | --- | --- |
|  | Total | % |
| **All restaurants** | **489** | **64** |
| *Sit-down dining* | 366 | 48 |
| *Fast-food* | 60 | 8 |
| *Buffet* | 22 | 3 |
| *Other or unknown* | 28 | 4 |
| *Multiple types* | 13 | 2 |

Source: CDC (2019)

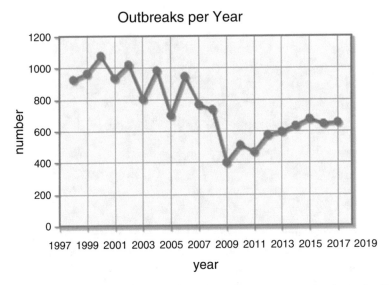

**Fig. 1** Total foodservice establishment-associated foodborne disease outbreaks in the United States between 1998 and 2017. (Data from the CDC National Outbreak Reporting System (NORS), https://wwwn.cdc.gov/norsdashboard/2020)

tions that are not able to be fully investigated, or are those with unknown cause but occurred in a foodservice establishment.

Nevertheless, when just looking at reported national foodborne disease outbreaks beginning from when the CDC first started tracking reported cases, total foodservice establishment-associated foodborne disease outbreaks appeared to peak in the year 2000 causing **1076** outbreaks, and then dropped to a low in 2009 of **398** (only once), and then increased each year up to 2017 with **653** cases of illness (see Fig. 1, data from CDC National Outbreak Reporting System (NORS)). Thus, although progress has been made, foodservice-associated foodborne disease outbreaks continue unabated, and if there was ever a time when the subject matter of this book (a paradigm shift to use Food Safety Management Systems (FSMS) to achieve Active Managerial Control of foodborne illness risk factors)) was needed it is NOW. No one should get sick, or worse, die from eating food simply because a business will not manage and prevent known risk factors associated with foodborne illness.

Historically, the contributing factors (see **Appendix A**) of the majority of foodborne disease outbreaks in foodservice establishments have been associated to just five foodborne illness risk factors originally described by the CDC (for more detailed information on these risk factors, see **Chap. 2**). The original report of the top five foodborne illness risk factors that were discovered as the cause of the majority of foodborne disease outbreaks from restaurants was reported by the CDC in its report *Centers for Disease Control and Prevention (CDC) Surveillance Report for 1988 – 1992* (CDC 1996). The five foodborne illness risk factors reported were directly related to food preparation processes within all foodservice establishments due to:

- **Food from unsafe sources** (e.g., receiving temperature and date of expiration/ proper storage compliance issues)
- **Poor personal hygiene** (e.g., employees working when sick with foodborne illness, and lack of barriers to cross contamination of foods by hands)
- **Inadequate cooking** (e.g., not cooking foods to the required temperatures that will kill all pathogens)
- **Improper holding/Time and temperature** (e.g., not holding foods at the proper temperature allowing for growth of bacterial pathogens and/or production of toxins)
- **Contaminated equipment/Protection from contamination** (e.g., not cleaning and sanitizing food contact surfaces, dishware, and food utensils, and the food-service environmental surfaces that lead transmission of pathogens from hands to foods)

The current means for the overall public health management of foodborne illness risk factors in restaurants have been via periodic health inspections by regulatory agencies and/or third-party audits by corporate multi-unit restaurant businesses (restaurant chains with two or more restaurants or franchised restaurants of the same concept often times located in one or more different states). Normally local health departments in each state perform health inspections of restaurant facilities and food preparation operations to measure how well a restaurant is following the requirements to prevent these risk factors associated with foodborne disease outbreaks (following a versions of the FDA risk-based inspections form; see FDA 2017a, Annex 5). Likewise, many corporate multi-unit restaurant businesses use third party auditors (normally every business quarter) to inspect their restaurants. These third party auditors observe corporate SOP's that may include quality, food safety, and customer experience requirements, and often times include specific audit of food safety food preparation procedures and their controls to prevent the top five risk factors associated with foodborne disease outbreaks; often modeled after the questions used in the FDA risk-based inspections form.

Periodic health inspections by regulatory authorities are necessary to protect the public health, and third-party audits are also helpful to ensure restaurant chain associated franchisee locations are following proper food safety procedures to make safe food specific to their recipes. There is significant evidence for the value of health inspections on the prevention of foodborne disease outbreaks. Evidence for the predictive value of health inspection data to the likelihood of future outbreaks in restaurants has also been demonstrated (for more detailed information, see Petran et al. 2012). However, these periodic health inspections/audits are forms of *passive management* where the audit is only a "snap-shot" of operations at one periodic event in time, and the corrective actions to ensure controls of hazards are re-established is also periodic (often times leading to their reoccurrence), and not always re-evaluated during another inspection/audit until months later. In order to consistently prevent foodborne disease outbreaks, foodservice establishments need to establish and execute, and regulatory agencies should inspect (see **Appendix A**) for the active management of the foodborne risk factor controls, where the assessment is daily, and corrective actions are performed in real-time before a hazard is introduced into a food product.

I believe a paradigm shift must now occur (see King 2016) across all foodservice businesses in partnership with state and local health agencies to execute daily **Active Managerial Control** of all foodborne illness risk factors using Food Safety Management Systems (FSMS). This change must occur if we are to significantly reduce, sustain this reduction, and one day prevent all foodservice-associated food-borne disease outbreaks in the United States. This paradigm shift should be in addition to and include current passive management systems (inspections and audits) discussed earlier because validation of the right FSMS should be performed by public health inspection. FDA defines Active Managerial Control or AMC as "the purposeful incorporation of specific actions or procedures by industry management into the operation of their business to attain control over foodborne illness risk factors". Active Managerial Control embodies a preventive rather than reactive approach to food safety through a continuous system of monitoring and verification (see FDA 2017b, Annex 4).

The foundation of Active Managerial Control is the design of a Process HACCP plan (hazard analysis and critical control point) and Prerequisite Program (see **Chap. 3**) that are then used to develop and execute FSMS (see Chap. 4). A hazard analysis must be performed to identify what hazards are associated with the food-service establishments food preparation processes and business, and controls must be defined. Process HACCP is very similar to HACCP used in the manufacture of human foods, but is designed for use in foodservice establishments (for a review of the comparisons of HACCP with Process HACCP see **Chap. 1**). The FSMS are then used to train employees and monitor food preparation processes and other controls in the Prerequisite Program to ensure each hazard is under a control, and managed by a Certified Food Protection Manager (CFPM).

A similar paradigm shift has already occurred in the human food manufacturing industry by the recently implemented Food Safety Modernization Act (FSMA) and the FDA enforcement of the Hazard Analysis and Risk Based Preventive Controls (HARPC) requirements detailed in these rules during human food manufacturing. Hazard Analysis and Risk Based Preventive Controls is part of the requirements of the FDA's Preventive Controls for Human Foods rules, which includes a hazard analysis and the active management of the controls required to prevent each defined chemical, biological, and/or physical hazard associated with any food manufacturing process and facility. In addition, documentation that these controls were effective at the time (active) of production must also be performed, and documents retained for inspection review.

To comply with the HARPC requirements, the controls necessary to prevent ingredient and facility related hazards must be defined by each food facility according to a hazard analysis of every food product and described within a product specific Food Safety Plan by a certified person called a Preventive Controls Qualified Individual (PCQI); this plan must outline where the controls of each hazard will be placed and monitored to ensure each is active, and must also describe what actions will be taken (corrective actions) when a control is not present. The HARPC requirement is basically a HACCP method that includes additional prerequisite program controls (like facility cleaning and sanitation, and environmental surface monitoring for pathogens), and monitoring/documentation requirements, and is very similar to foodservice Process HACCP. The HARPC rules provide value to the foodservice and retail food sales industry by helping ensure, in part, the ingredients and food products they source are safe; one of the top five risk factors (Foods from unsafe sources, see above). The methods used to establish supplier food safety and safe source of foods are addressed in numerous publications elsewhere, and we describe in another book with my colleague Dr. Wendy Bedale how to use HARPC methods to ensure each ingredient and product sourced for a foodservice business is safe (King and Bedale 2017). Ensuring each food manufacturer/supplier is controlling all hazards via HACRP is critical to the prevention of foodborne disease outbreaks in foodservice businesses; it serves as a critical component of the FSMS to establish Active Managerial Control in foodservice (also discussed in more detail in Chap. 4).

In my first book of this Springer series on Food Microbiology and Food Safety, Practical Approaches, called Food Safety Management (King 2013), I discussed the importance of supply chain and restaurant management systems, and their role in the design and execution of a food safety management program for restaurants to prevent foodborne illnesses; these as part of an effective corporate multi-unit restaurant business (and useful for any business that performs foodservice processes). This included both passive and active verification methods that a business could implement to ensure effective food safety controls were functioning across multiple supplier facilities and multiple restaurant locations from the perspective of a corporate food safety management program leader.

In this second book written on food safety management in this series, I want to share a more detailed perspective at the individual restaurant operations level (who are tasked to daily control the hazards associated with food preparation and service), and discuss how to implement FSMS that will achieve Active Managerial Control of the top five risk factors in a restaurant business. The 32 process failures that the CDC defines as contributing factors to the cause of foodborne illnesses (that lead to the risk factors, see Appendix A) can be prevented in a foodservice business via the design and execution of FSMS. The value of using these FSMS go beyond prevention of foodborne diseases (see **Chap. 8**), and are proven to also enhance food quality, restaurant cleanliness, customer experience, employee performance and health, and directly increase sales (and reduce cost) in a foodservice business (trust me, I have observed this). I will often interchange the words "restaurant," "foodservice," "retail foodservice business," or "retail sales with foodservice" business in this book because foodservice is now performed in multiple businesses including even retail convenience stores and gas stations. Foodservice is also performed in colleges/universities/schools, hotel/motels, catering businesses, banquet facilities, festivals/fairs, and even food trucks, each of which will benefit from using Process HACCP-based FSMS in their business.

The foodservice industry (as does the retail food industry) can have the most direct influence on the safety of our food supply in the United States across the network of food manufacturing as the buyer that can determine specifications for all food ingredients and products they purchase and will accept upon receiving. This includes influence over their food manufacturers/suppliers, their food distribution and storage business partners, and of course during the inspection and accept/reject actions of food delivered to their foodservice establishments. Local and state regulatory agencies are also critical to ensure oversight and accountability to the preparation of safe food in foodservice businesses. The CDC, FDA, and USDA have a significant role in defining hazards and guidance on the best means for control of foodborne illness hazards in the food supply chain, and the FDA, USDA, and many state agriculture agencies provide accountability by inspecting food manufacturing facilities to enforce and provide oversight for the production of safe food. However, neither the local health department nor government regulatory agencies can be in a food manufacturing nor foodservice facility every day, respectively. Thus the foodservice business has the most direct control of foodborne illness risk factors (like a final gate) before the food is served to consumers.

The collaborations to help identify new hazards and innovate improvement of the controls should continue to be fostered by partnerships between the industry, academia (by whom the peer-reviewed science for preventive controls is primarily described), government, and industry and government trade organizations. This should be done in advance of the local, state, and federal government's need to define updates to regulatory requirements (see the **Foreword** to this book by my friend and colleague of many years – who also led the food safety management program for a large corporate multi-unit restaurant – and who we both shared many best practices with each other over the years to improve our respective businesses even as or companies were competitors in the foodservice industry). Likewise, other food safety professionals in the foodservice industry can also share best practices for improved controls that they have successfully practiced in their businesses, as food safety knowledge is not a competitive advantage but a public health requirement and "the right thing to do". I firmly believe that when foodservice businesses can experience the value of implementing FSMS to their bottom line and in partnership with regulatory agencies, the outcomes of fewer foodborne illnesses and outbreaks can be achieved and sustained across the United States.

Saint simons Island, GA, USA                                                          Hal King, Ph.D.

# References

Centers for Disease Control and Prevention (CDC) (1996) CDC surveillance summaries. October 25, 1996. MMWR 1996.45 (No. SS-5)

Centers for Disease Control and Prevention (CDC) (2017) Surveillance for foodborne disease outbreaks, United States, 2015. Annual report. US Department of Health and Human Services, Atlanta, Georgia

Food and Drug Administration (FDA) (2017a) Annex 5. Conducting risk-based inspections. FDA Food Code

Food and Drug Administration (FDA) (2017b) Annex 4. Management of food safety practices-achieving active managerial control of foodborne illness risk factors. FDA Food Code

King H (2013) Food safety management: implementing a food safety program in a food retail business. Springer

King H (2016) Implementing active managerial control principles in a retail food business. Food Safety Magazine, February/March. Available at http://www.foodsafetymagazine.com/magazine-archive1/februarymarch-2016/implementing-active-managerial-control-principles-in-a-retail-food-business/

King H, Bedale W (2017). Hazard analysis and risk-based preventive controls: improving food safety in human food manufacturing for food businesses. Elsevier

Petran RL, White BW, Hedberg CW (2012) Health department inspection criteria more likely to be associated with outbreak restaurants in Minnesota. J Food Prot 75:2007–2015

# Contents

# Abbreviations

| | |
|---|---|
| AMC | Active managerial control |
| ANSI | American National Standards Institute |
| CDC | Centers for Disease Control and Prevention |
| CFPM | Certified food protection manager |
| CIFOR | Council to Improve Foodborne Outbreak Response |
| CAC | Codex Alimentarius Commission |
| CFP | Conference for Food Protection |
| CCP | Critical control point |
| EPA | Environmental Protection Agency |
| FMEA | Failure mode and effect analysis |
| FIFO | First in first out |
| FAO | Food and Agriculture Organization |
| FDA | Food and Drug Administration |
| FCS | Food contact surface |
| FSMS | Food safety management system(s) |
| FSMA | Food Safety Modernization Act |
| GM | General manager |
| GFSI | Global Food Safety Initiative |
| GMP | Good manufacturing practices |
| GRP | Good retail practices |
| HACCP | Hazard analysis and critical control point |
| HARPC | Hazard analysis and risk based preventive controls |
| IFSAC | Interagency Food Safety Analytics Collaboration |
| IAFP | International Association of Food Protection |
| ISO | International Organization for Standards |
| IOT | Internet of things |
| LTO | Limited time offer |
| NASA | National Aeronautics and Space Administration |
| NACMCF | National Advisory Committee on Microbiology Criteria for Foods |
| NEARS | National Environmental Assessment Reporting System |
| NSF | National Sanitation Foundation |

| OSHA | Occupational Safety and Health Association |
| PIC | Person in charge |
| PPE | Personal protection equipment |
| PHF | Potentially hazardous food |
| PCP | Prerequisite control point |
| QAC | Quaternary ammonium compounds |
| RFID | Radio frequency identification |
| RTE | Ready to eat |
| SDS | Safety data sheet |
| SOP | Standard operating procedure(s) |
| TCS | Time/temperature control for safety |
| USDA | United States Department of Agriculture |
| US | United States |
| WGS | Whole genome sequencing |
| WHO | World Health Organization |

# Chapter 1
# Introduction

*Repetition yields constants*
*Constants create cultures*
S. Truett Cathy

## Is It Not a Question of IF but WHEN?

Have your customers complained to you or posted negative complaints on social media against your foodservice establishments (or those franchised foodservice establishments that operate under your corporate business) relating to foodborne illnesses like vomiting or diarrhea or just not feeling well, strange taste/smell of a food product or an allergic reaction to a food, or a small piece of plastic or metal in a product? Have your foodservice establishments frequently discovered product defects (or customers return food with them) in the source of food ingredients you source from suppliers while preparing the food, and/or do you experience constant food ingredient/product withdrawals or recalls that force you to reorder replacement products? Do your foodservice establishments continue to commit the same repeat critical violations on health department inspections, earning poor inspection scores/grades? Have you avoided eating in a foodservice establishments due to cleanliness issues? If you have one or more of these conditions in your foodservice establishment(s), your current strategy for managing food safety risk, and likely ability to prevent foodborne illnesses is not working, and you need this book.

If you search via Google the search terms "food poisoning," "foodborne illness," "diarrhea," and "vomiting" along with any restaurant brand name in the United States (just use the word "AND" between the words), you will see that you are not alone. According to the recently published data by the US Centers for Disease Control and Prevention (CDC), restaurants continue to be the leading cause of foodborne disease outbreaks in the United States, causing 64% or more of outbreaks *each year* from 1998 to 2017 (CDC 2019). How can you know if you are at risk of causing (or already have caused) a foodborne illness in your customers? Foodservice business owner/operators can do something they already do which is to listen to

© Springer Nature Switzerland AG 2020
H. King, *Food Safety Management Systems*, Food Microbiology and Food Safety, https://doi.org/10.1007/978-3-030-44735-9_1

**Table 1.1** Example of relating two or more customer complaints of similar illness type reported on the same day of purchase (via phone call, social media post, etc.) due to biological hazards with possibility of association to a foodborne illness requiring further investigation by the foodservice operator. Not a complete list, shown as example only. For a complete list of all biological, chemical, and physical hazards, the predominant signs and symptoms of disease, their incubation periods, and the likely foods involved, see International Association for Food Protection (2011)

| Customer complaint-predominating signs and symptoms of illness type | Purchased food on | Biological hazards | | Example of likely menu items |
|---|---|---|---|---|
| | | Average time until illness after eating the food | Etiologic agent and source (biological hazard) | |
| Nausea/vomiting | Same day as complaint | 5 h | Exo- enterotoxin from *Bacillus cereus* | Fried rice/pasta, corn-meal bread |
| Abdominal cramps/diarrhea | Same day as complaint | 8–22 h | *Clostridium perfringens* endo-enterotoxin | Cooked meats, poultry/gravy/sauces, soups/refried beans |
| Fever/chills/malaise | Same day as complaint | 16 h | *Vibrio vulnificus* | Raw oysters and clams |
| Visual impairment/tingling and/or paralysis | Same day or next as complaint typically | 18–36 h | *Clostridium botulinum* (botulism) | Canned low-acid foods/smoked fish/cooked potatoes/onions/garlic in oil/frozen pot pies |

their customers and determine if they have the risk factors that are known to contribute to foodborne illnesses to know. For example, if you experience this (using the information in Table 1.1 to show criteria):

- Two or more customer complaints of illness with the same symptoms (e.g., *abdominal cramps/diarrhea*)
- About the same menu item of which you know you prepared and served that day (e.g., *beans*)
- Each stating they got sick the same day that they purchased and ate the food, and the customers have evidence that they purchased the food (*purchased food on same day they got sick*)
- And your restaurant cannot show it had these hazards under control at the day of the food preparation and service (e.g., *cooked the beans properly and held them at the proper hot holding temperature*)

these customer claims of illness are likely valid and could be caused by the biological hazard *Clostridium perfringens*. Remember that these are not the only hazards in restaurant foods as there are numerous biological, chemical (e.g., allergens), and physical hazards (see below) that can cause a foodborne illness or injury. Thus, no customer complaint of illness or injury should be ignored, and each should be investigated.

## What Can Be Done?

The most important means to proactively ensure food safety in a retail foodservice business is to establish and follow a food safety management program that focuses on continuously identifying and preventing all hazards. A food safety management program, whether established in an independent restaurant or one in a multi-unit and/or franchised restaurant chain, should have multiple Food Safety Management Systems (FSMS—see below) and trained personnel (i.e., the owner, manager, and employee are knowledgeable of food safety risk and controls), who are empowered to manage processes, and conduct and monitor specific risk-based preventive controls. These personnel will proactively work in the business to prevent hazards in the supply chain and during retail foodservice operations (King 2013, 2016). To most effectively prevent foodborne illness at a foodservice establishment operation, the business must also achieve daily Active Managerial Control (AMC) of the risk factors that contribute to most foodborne illnesses and disease outbreaks in the United States (FDA 2017a, b).

## What Is Active Managerial Control and Why Is It Needed Now?

According to the FDA Food Code (2017a, b), Active Managerial Control (AMC) of foodborne disease risk factors can be defined as the purposeful incorporation of specific actions or procedures by foodservice/restaurant management into the operation of their business to attain daily preventive controls over foodborne illness risk factors. Basically the "Active" in AMC is the continuous work to define the hazards, establish their effective controls, and then ensure each control is actively monitored during the time of all food preparation processes. In order to define the hazards and their preventive controls, a foodservice/restaurant business must use a method called Process HACCP (Hazard Analysis and Critical Control Point; FDA 2017a, b), and then design FSMS based on a Process HACCP plan that will be used to train employees, monitor the controls of each hazard, and provide immediate corrective actions during foodservice operations. Achieving AMC establishes a preventive FSMS that provides a proactive—rather than reactive—approach to food safety management as opposed to periodic audits/inspections that are mainly reactive (see below). The use of Process HACCP in retail foodservice establishments is based on HACCP principles (to adapt HACCP applications in non-food processing environments), where identified biological (e.g., *Salmonella* or *Norovirus*), chemical (e.g., a pesticide or allergen), and physical (e.g., a bone in a chicken nugget or piece of metal in a soup) hazards are placed under daily management to eliminate these hazards in food preparation processes that include receiving, storing, preparing, cooking, cooling, reheating, hot holding, and serving foods.

HACCP is a well-known method for reducing the risk of foodborne illness from foods made in human food manufacturing facilities (see Chap. 3 for more information about HACCP). Although Process HACCP is similar (HACCP is foundational) to HACCP, it includes more detail in the development of a plan in foodservice (Fig. 1.1; see Chap.3 for detailed method). HACCP is more easily developed and implemented in a food manufacturing facility where the facility produces one food product at a time on each individual production line (i.e., where the "flow of food" preparation is contained in one continuous line, and the HACCP plan describes controls). In contrast, restaurants and other retail foodservice establishments produce multiple food products (i.e., the menu) within the same food preparation area (e.g., a kitchen often with no defined area for raw vs. RTE food prep and/or proper flow of food design—see Chap. 6). Again, the FDA Food Code states that the best means to accommodate this difference is to apply HACCP to a retail foodservice establishment (see Annex 4, FDA Food Code, FDA 2017a) via what it terms as Process HACCP.

The FDA Food Code gives more details on the primary differences between retail foodservice operations, and those operations found in food manufacturing (FDA 2017a, b) to help better explain why Process HACCP is more effective. Unlike food manufacturing businesses such as canning, dairy, or produce processing facilities, most retail foodservice facilities are not easily defined by a single commodity or process. Similarly, most foodservice establishments have the following attributes that clearly differ from most food manufacturing facilities:

- High employee and manager turnover, exceptionally high with entry-level employees

  - The employees and managers have very little experience in their job duties; continuous food safety training is required for each new employee throughout the year which is often not performed due to high turnover rates and cost to train.

- Many foodservice businesses are start-up companies or newly franchised businesses in which the owners were not previously in the food preparation and sales business.

  - They operate as single-location businesses without the benefit of corporate-level food safety training or FSMS designed specifically for their menu.
  - They have low profit margins; thus labor and time costs are carefully managed, reducing time available for an employee's food safety training.
  - They renovate existing buildings not originally designed to accommodate kitchens.

- Many foodservice businesses have multiple, constantly changing, menu items (e.g., many Limited Time Offers (LTOs)), recipes, new products, sourced ingredients, processes, methods, and production volume changes (e.g., coupons or dollar menu items to increase sales) each day/season; their workflows are not easily adapted to one standard operating method without a Process HACCP plan.

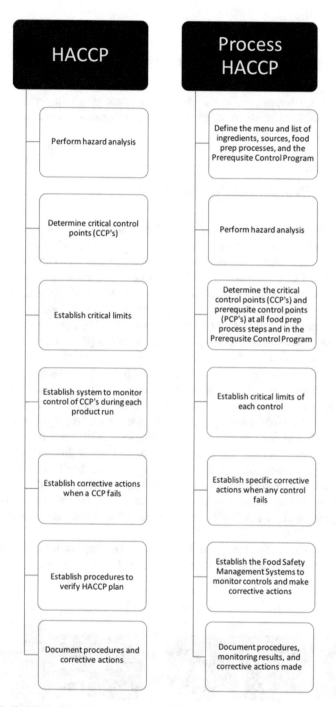

**Fig. 1.1** Similarities in HACCP used in food manufacturing compared to Process HACCP recommended for retail foodservice establishments. Hazards and their controls are defined according to ingredients used and food preparation processes performed to produce a human food product. Management of the controls during food preparation processes, and in the Prerequisite Control Program is performed daily to prevent the hazards using Food Safety Management Systems

- – Changes occur on a daily basis with little to no notice or time for preparation (e.g., a large same-day catering order when understaffed).

- • Customers enter, eat, and use the dining rooms and restrooms in the facility close to areas of food production; employees have common use of customer-facing facilities.

## How Does Process HACCP Work?

Process HACCP is the method of first defining the flow of food preparation in a kitchen (like the flow of food in a manufacturing facility production line) for a restaurant's menu, determining the hazards at each process, then establishing the controls (CCP or PCP; see below). All foodservice establishments have eight primary food preparation processes in common based on stages (activities) in their food preparation of their complete menu (establishing a "flow of food" from when food is received into the restaurant through when it is served to a customer; see Fig. 1.2). Some foodservice businesses may not perform every process (e.g., may not cool down and then reheat foods), but all will follow the same flow of food and oftentimes perform additional activities or stop some based on the recipe and menu (e.g., LTOs may require cook, cool, and reheat processes not normally used in the regular food preparation plans for the menu) or the business may delete menu items.

Most of the hazards are already well known at each of the food processes in this flow of food, and are common to all foodservice establishments, only differing based on the ingredients used to prepare food (e.g., raw ground beef ingredients have a Shiga-toxin-producing *E. coli* hazard associated with them). Controlling these hazards at each food preparation process (and during other activities necessary to the business; see below) can achieve the same risk reduction as if the business were to prepare an HACCP plan for each individual menu item (for more details, see Chap. 3). Not all controls in Process HACCP are placed as Critical Control Points (CCPs) in the food preparation processes, but may include control measures necessary in the Prerequisite Control Program. I call these Prerequisite Control

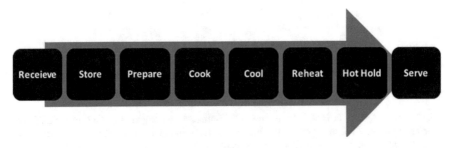

**Fig. 1.2** The flow of food preparation processes common in all foodservice establishments used to develop a Process HACCP plan based on the menu of the business. (Source: FDA 2017a, b)

Points (PCPs) to differentiate them from CCPs, and they include controls such as physical barriers (e.g., gloves to ensure no bare-hand contact of RTE foods), personal hygiene standards (e.g., a health policy and wellness checks to exclude/restrict sick employees from preparing food), and/or food safety procedures/tools to control hazards from cross-contamination of surfaces (e.g., using a cleaning and sanitation management system to reduce the risk of Norovirus transmission to surfaces).

Prerequisite Control Points (PCPs) are derived from a predefined Prerequisite Control Program which are activities not always directly associated with the food preparation processes of the business, but may introduce hazards (e.g., cross-contact of undeclared allergens in the kitchen) or require the control of hazards outside of the processes (e.g., supplier control of Shiga-toxin-producing *E. coli* in raw ground beef ingredients before distribution of the ingredient to the foodservice business). As discussed in Chaps. 3 and 4, when a Prerequisite Control Program is defined and executed with the Process HACCP plan, the FSMS are more effective and more comprehensive in controlling all of the hazards associated with the menu a foodservice business prepares and sells. While the Process HACCP plan is developed to identify hazards (via the hazard analysis) and their controls, the Prerequisite Control Program is primarily designed to identify known controls of contributing factors necessary to ensure the safe preparation of food. Many of the PCPs (like CCP's) derived from the Prerequisite Control Program are best defined and monitored during the food preparation processes where the hazards are most likely to occur (Fig. 1.3). This is discussed in more detail in Chaps. 3 and 4. The minimum recommended components of a Prerequisite Control Program necessary for effective FSMS include:

- Methods to ensure equipment maintenance:

  - Equipment to ensure temperature measuring devices are calibrated.
  - Cooking equipment is calibrated and hot and cold holding equipment provides the correct temperature.
  - Refrigeration and freezer equipment provides the correct environmental temperature.
  - Ware washing equipment are operating according to manufacturer's specifications.

- Methods to ensure allergen management in food preparation and storage
- Methods to ensure safe chemical use and storage around foods, and employee safety that meet Occupational Safety and Health Administration (OSHA) requirements for foodservice businesses (see https://www.osha.gov/SLTC/restaurant/).
- Methods to ensure safe water use for food and in the manufacture of ice
- An effective pest prevention program to prevent pest infestations
- Methods to ensure no bare-hand contact with any ready-to-eat (RTE) food to prevent the cross-contamination of foods from hands
- Methods to ensure proper hand washing to prevent the cross-contamination of foods from hands, including when wearing gloves

**RECEIVE**
*Known source **(PCP)**, receiving temperature **(CCP)***

**STORE**
*Proper storage temperature **(CCP)**, prevent cross contamination **(PCP)**, store away from chemicals **(PCP)**, no use past expiration date **(PCP)**, clean and sanitary storage containers **(PCP)***

**PREPARE**
*Personal hygiene **(PCP)**, restrict sick employees **(PCP)**, prevent cross contamination **(PCP)**, no bare hand contact with RTE foods **(PCP)**, clean and sanitary equipment and utensils, containers **(PCP)***

**HOLD**
*Critical limit of internal temperature and time of product **(CCP)**, prevent cross contamination **(PCP)**, clean and sanitary storage containers **(PCP)***

**COOK**
*Critical limit of internal temperature and time of product **(CCP)***

**COOL**
*Critical limit of internal temperature and time of product **(CCP)**, clean and sanitary storage containers **(PCP)***

**REHEAT**
*Critical limit of internal temperature and time of product **(CCP)***

**HOT HOLD**
*Critical limit of internal temperature and time of product **(CCP)**, prevent cross contamination **(PCP)**, clean and sanitary storage containers **(PCP)***

**SERVE**
*Personal hygiene **(PCP)**, restrict sick employees **(PCP)**, prevent cross contamination **(PCP)**, no bare hand contact with RTE foods **(PCP)***

**Fig. 1.3** Example use of PCPs defined in the Prerequisite Control Program and CCPs defined in the Process HACCP plan. Process HACCP and the Prerequisite Control Program together define the hazards and their priority control measures that must be performed during each step of each food preparation process (Source in part: FDA 2017a, b). Some PCPs may be assessed outside of the food preparation processes (not shown). RTE is ready-to-eat

- Methods to ensure restriction and exclusion of sick employees who have known signs, symptoms, or diagnosis of foodborne illnesses to prevent the cross-contamination of foods from hands including cuts and burns (as part of the health policy—see Chap.4)
- Personal hygiene requirements of employees (clean clothing, hair restraints, eating/smoking/drinking restrictions, jewelry restrictions)
- A cleaning and sanitation program for the direct prevention of cross-contamination of ready-to-eat foods by raw animal foods, clean and sanitized food contact surfaces, cutting boards, dish washing equipment, utensils, aprons, etc.

- Methods to ensure safe source of foods—using only food safe ingredient and food suppliers (e.g., GFSI- and FSMA-compliant supplier food manufacturing facilities) to ensure a safe source of food and food packaging that is safe to serve food on (e.g., meets FDA food packaging in contact with food requirements)
- Methods to ensure ingredients are not used past their safe expiration date using FIFO
- Methods to ensure an ingredient/food product in use has not been recalled by the FDA nor by a CDC "do not consume" communications to ensure the restaurant is alerted to when the FDA communicates not to serve an ingredient/food product or the CDC communicates not to consume an ingredient/food product

In fact, these last three bullets (FIFO and use of approved source of foods/packaging) in a Prerequisite Control Program will manage a critical foodborne illness risk factor (unsafe ingredient and food sources) that when not under control can increase the number of hazards in all the other food preparation processes (Fig. 1.2); this has led to many foodborne disease outbreaks from retail foodservice establishments even when the source of food was safe when ordered but became unsafe after it was recalled, and the business was not aware of the recall. Many multi-unit retail foodservice businesses have established supply chain management programs to ensure that food ingredients are safe. Products and packaging used by their establishments are safe according to national and international certification standards (e.g., Global Food Safety Initiative (GFSI)-certified food manufacturing facilities) or are purchased from broad-line food distributors that follow the same standards for supplier selection and distribution.

Some foodservice businesses may not be sufficiently established to develop a supply chain Food Safety Management System to ensure a comprehensive Prerequisite Control Program. For example, many new restaurants do not have a supply chain program to ensure sourcing of safe foods. However, recent enhancements to FDA regulations of manufactured human foods under the Food Safety Modernization Act (FSMA) are improving the means for retail foodservice establishments and corporate businesses to ensure safe food sources. The act mandates that food manufacturers adhere to the use of FSMS based on HACCP plus well-defined Prerequisite Control Programs (called Hazard Analysis and Risk-Based Preventive Controls) to achieve Active Managerial Control of manufactured food risk; very similar to a foodservice Prerequisite Control Program and Process HACCP. The Hazard Analysis and Risk-Based Preventive Controls method includes a requirement to define a food safety plan for every food product manufactured in a facility. Foodservice businesses can use these FDA rules (requesting the food safety plan for their prospective ingredients/products they wish to purchase) as a means to ensure safe source of foods including those with established supply chain Food Safety Management Systems discussed in more detail in Chap. 3 (also see King and Ades (2015) and King and Bedale (2017)).

## A Case Study: One Incident, Two Different Outcomes

Let's look at a simulated case study in a restaurant to discover how achieving AMC can make a significant impact on the management of hazards, resulting in prevention of a foodborne disease outbreak.

A restaurant manager, Betty, was cleaning a pot in the three-compartment sink in the kitchen of her new franchised barbecue restaurant. Suddenly, she noticed a strange woman walking through the kitchen talking to her employees. Betty grabbed a paper towel to dry her hands, and walked quickly over to intercept the woman.

The woman said "Miss, my name is Martha, and I am the county environmental health officer. I am here to investigate a foodborne disease outbreak that your restaurant may have caused this week. As of today, we have over 400 people who have reported to the health department or went to Mercy Community Hospital because of constant vomiting, diarrhea, and flu-like symptoms for the last 2 days. All of these folks claim that they ate at your restaurant this week or had food you prepared and catered to a party. One person who had similar symptoms has died, though she resided in a family care center for the elderly—we're not sure if she is a related case."

"What? Uh, you must be kidding, right?" squeaked Betty.

"No, ma'am," Martha said, "this looks like a *Norovirus* outbreak due to the rapid onset of illness—an open-and-shut-case, I think." Martha began walking toward the hot food prep line, with Betty following closely behind to watch what she was doing.

"I just don't know how this could have happened," Betty said, "I'm ServSafe certified, and we cook all of our food; it's barbecue, how can nervo-virus grow in barbecue?"

"Well, ma'am," Martha replied, "it's a virus that doesn't grow in barbecue like bacteria can grow, but it sure can *contaminate* the barbecue if one of your employees happened to be sick, or if you had bad produce in your salads, or used other unsafe food sources. It is very difficult to stop the spread of the virus if you don't ensure that any sick employees do not handle food, if you don't have proper hand washing and/or glove use requirements when employees are working with ready-to-eat foods like barbecue, or if you don't carefully ensure that your ingredients come from reputable sources. Most of the time, it's an undetected sick employee who forgets to wash their hands and/or not wearing single-use foodservice gloves before handling RTE food, but it can also come into the restaurant on produce or seafood."

As Martha walked through the restaurant, she asked Betty if she knew of anyone who was out sick or came to work sick or any employee or customer who might have gotten sick like vomiting in the dining room or in the restroom. She also asked to see Betty's health policy. "We don't know if there are more sick people because some people don't report their illness to the health department or visit their doctor," Martha noted. "We likely will see more cases reported this and next week especially as folks go to the hospital if this is in fact a *Norovirus* outbreak."

## *Two Different Outcomes Based on the Absence and Presence of AMC*

### Scenario 1: Without AMC

Betty said, "I remember from my ServSafe training that I needed to ask employees if they were sick and, if so, tell them not to work. But no one looked sick this week and we were short of staff—we really needed 'all hands on deck', so to speak, with this holiday rush." Betty frowned and thought for a moment. "Josh, the stock guy, did spend a lot of time in the restroom, but he said he was keeping it cleaned up regularly. I use Josh a lot on busy days to do 'fill-in' jobs when we get busy." Martha looked dubious.

Martha looked around the kitchen and asked, "Do you require gloves when employees like Josh prepare and handle ready-to-eat foods like those barbecue sandwiches? Did Josh do any food prep this week?"

"Uh, no and yeah," Betty affirmed, "Josh fills in during the busy parts of the day, like I said. He helped make barbecue sandwiches on the prep line twice this week because we had large catering orders."

Martha asked, "Do you have a hand washing requirement before they do that type of work—and can you validate this?"

Betty answered, "We talk about hand washing all the time, and we have a third-party audit four times a year that corporate does to validate it. We don't allow employees to wear gloves because the employees don't wear them properly when they change tasks, and we aren't required by corporate to wear them."

Martha asked, "Do you have a method to ensure that employees cook all the food at the right temperatures?" Martha checked the temperature of the pot of Brunswick stew cooking on the stovetop, and took some notes on her notepad. Then she asked, "Can you show me the records of your produce vendors? Some of the sick people said they ate your barbecue salad, too."

Betty looked confused. "We don't have records, just receipts if that is what you're asking" she said. "The produce distributor helps find us the lowest-cost produce so that we can reduce our food cost. Produce goes bad quickly, you know."

"I will need to see all your produce purchases and sources over the last few weeks," Martha said, looking concerned.

Martha asked Betty to join her in the office. Betty followed Martha into the office with a mounting sense of dread.

When Betty shut the office door, Martha gave her the bad news. "Betty, I would recommend that you close the restaurant, at least for the rest of the week. Ensure that all of your employees are healthy and do a deep clean and disinfection of your entire restaurant—especially all food contact and non-food contact surfaces (e.g., knobs on cook equipment, hand wash sinks, etc.) and restrooms (door handles, stall latches, sink faucet handles, etc.), food storage, and any other surfaces that employees touch or that come into contact with food."

Martha continued, "I will have to report this in my inspection. I'm sorry to say that it likely will be picked up by the media, too." Martha handed the dejected Betty a list of actions to be taken immediately before reopening the restaurant. "These will help ensure that you prevent any further cases in this outbreak before you open back up," Martha concluded, "I hope that lady that died due to Norovirus didn't eat here, we haven't been able to determine this yet."

**Scenario 2: With AMC**

Betty said, "We have a health policy that we use to train employees that also requires employees to report to us when they get sick. I, along with my other managers, do a wellness check with each employee when they check in to work, and in the middle of each shift. It's really pretty simple; we just ask them if they have had any symptoms like vomiting, fever, or diarrhea. Also, if we notice anything unusual, we may restrict employees from food-prep work, even if they say they aren't sick just to be safe like if they have bad cuts on their hands or arms."

"We also keep a sick log; no one has been on the log for the past four weeks," said Betty, showing Martha the log. "We did have a few out with the flu last month, though."

Betty continued, "No customers or employees became ill in the dining room or restrooms either. Oh, and I sent Josh home immediately a few months back, when I noticed that he was spending too much time in the restroom. When he came back after a few days, I didn't allow him to work with food for the rest of the week. He didn't have any of the signs or symptoms you mentioned each day we did our wellness check, or any of the disease names like *Norovirus* or *E. coli* O157 that we track and ask about each day." Betty added, "Josh is a good worker. The only reason he is not a manager here is because he is in school and can work only part-time. He even has his ServSafe certificate!"

Martha looked around the kitchen and asked, "Do you require gloves when employees like Josh prepare and handle ready-to-eat food like those barbecue sandwiches? Do you have a hand washing requirement before they do that type of work?"

Betty answered, "We require regular hand washing before any food prep, and our managers monitor this during the preparation of foods. All employees are also required to wear color-coded gloves: for example, everyone knows that they are allowed to work only with ready-to-eat foods when they have blue gloves on. In fact, we record this every day using our daily food safety self-assessment you folks call a Food Safety Management System. We monitor things like 'blue glove use only at the ready-to-eat food prep stations' and watch employees washing their hands to ensure that they are always doing it right before food prep."

Betty continued, "I am very confident that even if an employee of mine was sick, and we didn't catch it with our self-assessments, the proper hand washing and glove use would have prevented it."

Martha nodded, taking note of the fact that the hand sink in the kitchen had not only soap and sanitizer but also a job aid to advise employees when and how to wash hands and remind them to use a sanitizer afterward for extra safety. "Always clean and sanitize hands" was the title on the sign.

Betty added, "We also require all employees to wash their hands twice after they go the restroom for any reason: once in the restroom and then again when they return to the kitchen before they do anything else. We monitor this too by watching employees when they come back into the kitchen. Everyone washes their hands when they come into the kitchen no matter what."

Martha asked, "Do you have a method to ensure employees are cooking all of the food at the right temperature?"

"Yeah," Betty affirmed, "we use a Bluetooth thermometer and a food-check mobile app on our smart phones to record temps after cooking every large batch of meats and stews, and every few hours at the warming line. We also check temps this way in the cooler, so we are certain that all food is cooked, cooled down, and held properly—and that we can prove it." Betty emphasized, "this mobile app has really helped us track and validate that our Food Safety Management Systems and each control in our Process HACCP plan is working. We use the Process HACCP method, and define our Prerequisite Control Program to ensure we cover all the bases."

"Betty, can you show me the records of your produce vendors? Some of the sick people said they ate your barbecue salad," Martha explained.

"Sure, we only use corporate approved brands for all our produce—as matter of fact, we do this for all our foods delivered to my restaurant," Betty clarified. "Corporate makes us check the most current approved supplier/vendors list and temps whenever we get deliveries. We have to reject the produce if it's not on the approved source list, and we have to report this to corporate. They do all the work to ensure that our produce suppliers and distributors meet all the food safety specs. We also make sure that we wash the produce twice: once before we prep it and then again after we prep it, both times in a separate, produce-only sink."

## After Events Analysis

You can quickly see in Scenario 1 that without a FSMS to achieve AMC, it is probable that all the evidence points to Betty's restaurant as the cause of the outbreak. Although the health department would continue its investigation to confirm the likely source of the outbreak, the lack of AMC in Betty's restaurant and presence of multiple foodborne disease risk factors makes her restaurant a high-risk establishment for this likely association to and/or future outbreaks of foodborne diseases.

However, in Scenario 2 with AMC assured by a FSMS described in part by Betty in this case study, (1) the likelihood of the illnesses being caused by her restaurant is very low; (2) if an employee did work while sick in her restaurant (e.g., an asymptomatic employee with *Norovirus* worked while sick), the degree of the outbreak would be limited to much fewer cases or not at all; and (3) it could help the investigator to more quickly determine the likely root cause where the real source of the outbreak was from if, for example, the customers also ate foods from other establishments during the same time period that did not have AMC in place.

## Why Daily Food Safety Management Systems vs Audits

Most restaurants in the United States are inspected by a local health department inspector in each state. This inspector performs an audit of the foodservice location based on a state version of the FDA Food Code risk-based inspection standards. These standards include both a check for the presence of the controls of foodborne illness risk factors (focused on the top five risk factors; see Chap. 2) and a check for the presence of good retail practices (GRPs are focused on things like cleaning and sanitation programs, pest control, etc.) that, together, should be present in all restaurants. These health inspections are normally graded/scored, and the business is required to post the grade/score/full report for public disclosure of food safety risk in that establishment. Many states also post the individual restaurant location grades/scores on a public website. These health inspections are an important public health tool used, in part, to ensure that restaurants are executing proper controls of foodborne illness risk factors. The results of the inspections can often be predictive of the probability of a restaurant causing a foodborne disease outbreak, as reported by Petran et al. (2012). The investigators in this study made a comparison of past foodborne disease outbreaks in the state of Minnesota health department inspections data, specifically the critical violations also called foodborne disease risk factors related to the top five risk factors in individual restaurants involved in the outbreak. They found that a large percentage of outbreaks could be correlated to prior critical violations specific to the contributing factor and associated pathogen. For example, restaurants that caused a Norovirus outbreak consistently had violations on their health inspection forms specific to failure to exclude sick employees from work.

Traditionally, multi-unit/chain restaurant companies use third-party audits performed by auditing firms of their restaurants (oftentimes franchisees/owners of the business/brand) to determine how each is executing these proper controls of food safety risk. They will also use these audits to evaluate other business metrics like customer experience, employee safety risk, facilities and equipment maintenance, etc., which sometimes are viewed as more important to the business than food safety SOP execution audits. Most of these third-party audits are unannounced and performed by trained food safety auditors who use their own food safety standards, some of which mimic a health department inspection based on the most current FDA risk-based inspection form, to grade/score the restaurant operations.

Health inspections are rarely performed more frequently than once per quarter in most states; restaurants are most often inspected only twice per year—unless a restaurant has been cited for significant compliance issues (critical violations) or the restaurant concept/menu dictates more frequent inspections due to higher risk. Likewise, business-driven third-party audits are generally performed every quarter; however, the cost associated with time required to audit the restaurant and the number of restaurants audited across a multi-unit company sometimes limits their use. Both forms of evaluation of a restaurant's ability to control food safety risk are single snapshots in time that are reported to the public/business retrospectively. Although each provides valuable insight on execution at the time of inspection/ audit, and can include coaching and recommended corrective actions during the time of the inspection/audit, they rarely address root causes. Further, they are not comprehensive because of the inspector's limited time in the restaurant, which is often driven by large numbers of daily restaurant inspections required of the individual health department. In fact, a single mid-sized county or local jurisdiction in the United States can have over 1000–5000 foodservice establishments that a local health department with 5–10 trained inspectors must inspect regularly 2–4 times per year. A recent publication by a Harvard student reported that these health inspections were less effective in discovering out-of-compliance food safety controls when the individual inspector performed too many audits in a single day (Harvard Business Review 2019).

Most importantly, these health inspections and third-party audits do not monitor and manage these risks each day, nor do they foster accountability of the management process, performed by trained Certified Food Protection Manager (CFPMs)— all of which are necessary to prevent foodborne disease illnesses caused by restaurants.

## Why Food Safety Management Systems Vs Checklist

Many companies design standard operating procedures (SOPs) for all recipes (e.g., how to prepare and cook a hamburger) and food safety-related procedures (e.g., how to clean and sanitize dishware or food contact surfaces); they generate checklists to monitor these SOPs. However, in my experience, most checklists are just that: a list of items to check that are most often *not* based on the control of foodborne illness hazards or checked only incompletely; rarely do checklists focus on controlling the actual food safety hazard or with a defined process to correct each out-of-compliance issue. Likewise, many such checklists direct the user to check only quality-related metrics, and do not focus on the priority risk factors known to cause most foodborne disease illnesses and outbreaks from retail foodservice establishments. In contrast the Active Managerial Control of foodborne illness risk factors will monitor the controls, and ensure they are in place daily; it is more than a checklist. The system is designed to monitor operations and make corrective actions, which also enables root-cause assessment of failures (e.g., insufficient employee training or equipment malfunction).

## How to Use This Book to Achieve Active Managerial Control of Foodborne Illness Risk Factors in Your Restaurant(s)

This book is organized into chapters based on the core components and value of FSMS, how to design them based on Process HACCP principles, and the training, facilities, and digital technology necessary to achieve Active Managerial Control in a restaurant location. Further, it is organized to include program management concepts for multi-unit restaurants (e.g., a corporate restaurant chain or multi-unit franchisee owner).

- **Chapter 1** (this Chapter) provides an introduction to the problem of foodborne disease illnesses and outbreaks caused by retail foodservice businesses in the United States, and how you can know if your foodservice business is causing any, and describes how and why Process HACCP plans, a Prerequisite Control Program, and daily use of FSMS (that can achieve Active Managerial Control of the foodborne illness risk factors) are necessary to prevent foodborne disease illnesses and outbreaks. This chapter includes a "with and without AMC" case study that may also be useful for facilitating discussions with corporate leadership for program support of the design and implementation of FSMS in franchised restaurants.
- **Chapter 2** describes the known risk factors and associated contributing factors that lead to the hazards in foods prepared in foodservice businesses. When these hazards are not controlled by FSMS, they lead to sporadic illnesses and foodborne disease outbreaks. In order to develop the most effective FSMS, you must also know where the hazards are most probable so they can be controlled during the specific food preparation process or in the Prerequisite Control Program.
- **Chapter 3** discusses how to develop a Process HACCP plan (based on the menu) and the Prerequisite Control Program to define your hazards and how to define the controls (CCPs and PCPs) to prevent the hazards. This chapter also describes how to stay updated for new and emerging hazards in foodservice, and how to determine where to best control a hazard in the Process HACCP plan and Prerequisite Control Program.
- **Chapter 4** explains how to build your FSMS based on the Process HACCP plan and the Prerequisite Control Program for the foodservice business. This includes the methods to activate the FSMS in the business and ensure all controls are monitored at the appropriate time during operations. This chapter also describes the different types of FSMS that are necessary (e.g., cleaning and sanitation FSMS to prevent *Norovirus* transmission to restaurant environmental surfaces) to ensure all hazards that may indirectly increase the risk of the contamination of food are controlled.
- **Chapter 5** describes the necessary training to ensure management of and compliance to the FSMS including demonstration of knowledge of the Process HACCP plan and Prerequisite Control Program. This chapter discusses how and why the owner/GM, managers, and all employees must have knowledge of the foodborne illness risk factors, how they are controlled, and how the FSMS works to prevent foodborne illnesses and outbreaks.

- **Chapter 6** describes the necessary foodservice facility design that will enable and sustain effective FSMS. The chapter includes discussion of the necessary design that will enable the proper flow of foods in the kitchen to reduce cross-contamination of biological hazards and cross-contact of allergens with RTE foods. This includes effective signage that informs employees on the proper place for different food preparation processes that facilitate efficient and effective workflow.
- **Chapter 7** discusses the digital technologies that will enable the most effective FSMS including execution and verification of each individual CCP and PCP in the Process HACCP plan and Prerequisite Control Program. Digital technology is necessary to ensure continuous monitoring using the proper assessments and standards (and how to perform the assessments), predefined corrective actions (and alerts where needed), real-time updating (e.g., as menu changes or employees change), and documenting of results to enable data analytics and reporting.
- **Chapter 8** explains the value proposition to the business in using FSMS to achieve Active Managerial Control—the bottom line for the business.
- **Appendix A** describes the CDC 32 contributing factors and their relationship to each of the top five foodborne illness risk factors in a foodservice business. Each contributing factor description explains how a food becomes unsafe due to biological, chemical, or physical contamination, or proliferation/amplification, or survival of pathogens.
- **Appendix B** enables the reader to gain more knowledge on the future of Active Managerial Control and FSMS from the regulator's point of view. Although, currently, the FDA Food Code discusses the use of Process HACCP plans and FSMS to achieve AMC as an option for operators (see Annex 4, FDA Food Code, FDA 2017a), this food safety method has significant support from state regulators and the FDA (and me, of course) as the future gold standard in food safety management for preventing foodborne disease illnesses and outbreaks in foodservice establishments. I believe it should be and will likely be moved from Annex 4 into future FDA Food Code rules and/or incorporated into state rules that are regulated via health inspections.

The **References** in the back of each chapter are mostly "how to" sources, and are heavy on citing Internet websites to aid the reader in immediately finding and using additional resources for FSMS development.

Because the science of food safety is always improving due to the work of food safety professionals in government, academia, and industry, this book cannot possibly cover all aspects of FSMS design, address how to achieve Active Managerial Control in all possible circumstances, or touch on all other possible solutions for reducing the risk of foodborne illnesses and outbreaks within a retail foodservice business. You must remain a student of the many excellent resources (peer-reviewed journals, books (including the publishers of this book series), trade journal articles, and peer-reviewed research reports/papers) written by experts in all areas of food safety science. You should also attend and participate (speak about your best practices) in the numerous national and international organizations that support the development and communication of food safety science via conferences (like the Food Safety Summit, GFSI conferences, International Association for Food Protection conference, etc.), exchanges, and food safety product vendor-sponsored best practice meetings.

Ultimately, the responsibility of food safety within an organization comes down to people and their desire—not to mention the organization's commitment to empower them—to "own" the responsibility of food safety. Preventing human suffering that can result from a foodborne illness makes food safety a public health responsibility of all foodservice businesses.

Moreover, the negative impact that any foodborne illness inevitably has on a company's brand—think Chi-Chi's or Jack in the Box—should motivate all decisions by a foodservice business. Untimely, the best means to ensure the right FSMS are designed for the foodservice business is by employing thought leaders in food safety who carefully stay abreast of the hazards, and can help build a food safety management program to manage risks across the enterprise that includes the supply chain, logistics, and all foodservice facility designs and operations (King 2013).

This book is intended to interpret current knowledge from the FDA and CDC, academia, and my experience working in and with many of the most successful and largest foodservice businesses in the United States into a practical "how to" develop and maintain an effective FSMS in a foodservice business. If you, as a food safety professional, would like to provide feedback and additional ideas/improvements for how a foodservice business can manage food safety in support of its public health responsibilities, I welcome such comments on thefoodsafetylab.com (a blog for food safety professionals to post/share their best practices). These inputs could be important to improve subsequent editions of this book.

Finally, I'll encourage you by repeating the quote from the beginning of this chapter, which relates to a company's ability to establish a culture of safety across the enterprise:

> Repetition yields constants.
> Constants create cultures. (S. Truett Cathy)

Unless your business implements a standard that can be trained to, implemented and executed, and monitored to ensure it is constant every day during operations replicated consistently across the enterprise, you will not achieve a culture of safety.

# References

Centers for Disease Control and Prevention (2019) Surveillance for foodborne disease outbreaks, United States, 2017, annual report. US Department of Health and Human Services, CDC, Atlanta

Food and Drug Administration (FDA) (2017a) FDA Food Code. U.S. Department of Health and Human Services, Public Health Service. https://www.fda.gov/Food/GuidanceRegulation/RetailFoodProtection/FoodCode/ucm595139.htm

Food and Drug Administration (FDA) (2017b) Annex 4. Management of food safety practices–achieving active managerial control of foodborne illness risk factors. FDA Food Code, Silver Spring

Harvard Business Review (2019) To improve food inspections, change the way they are scheduled. See: https://hbr.org/2019/05/to-improve-food-inspections-change-the-way-theyre-scheduled

International Association for Food Protection (2011) Procedures to investigate foodborne illness, 6th edn. Springer, New York

King H (2013) Food safety management: implementing a food safety program in a food retail business. Springer, New York

King H (2016) Implementing active managerial control principles in a retail food business. Food Safety Magazine, February/March.http://www.foodsafetymagazine.com/magazine-archive1/februarymarch-2016/implementing-active-managerial-control-principles-in-a-retail-food-business/

King H, AdesG (2015) Hazard analysis and risk-based preventive controls (HARPC): the new GMP for food manufacturing. Food Safety Magazine, October/November.http://www.food-safetymagazine.com/magazine-archive1/octobernovember-2015/hazard-analysis-and-risk-based-preventive-controls-harpc-the-new-gmp-for-food-manufacturing/

King H, Bedale W (2017) Hazard analysis and risk-based preventive controls: improving food safety in human food manufacturing for food businesses. Elsevier, San Diego

Petran RL, White BW, Hedberg CW (2012) Health department inspection criteria more likely to be associated with outbreak restaurants in Minnesota. J Food Prot 75:2007–2015

# Chapter 2
# Hazards and Their Contributing Factors to Foodborne Illness Risk in Foodservice Establishments

*Foodborne illness is a significant problem in the United States. An estimated 48 million foodborne illnesses occur annually in the United States, resulting in approximately 128,000 hospitalizations and 3,000 deaths. Only a portion of these illnesses are associated with outbreaks (defined as two or more cases of a similar illness resulting from ingestion of a common food...).*
Brown et al. (2017)

## Foodborne Illnesses in Foodservice Establishments—Sporadic Events vs. Outbreaks

A foodborne illness is primarily caused by three types of hazards, two of which lead to illness and/or disease and the third of which causes physical injury; they are defined more specifically as:

1. Biological (e.g., *illnesses*, often called diseases, caused by microbial pathogens, toxins, prions)
2. Chemical (e.g., *illnesses* caused by allergens, pesticides, cleaning agents, heavy metals)
3. Physical (e.g., *injuries* caused by bones, plastic, metal shavings, glass, stones)

Biological hazards cause the most serious foodborne illness outbreaks because most of the agents are living organisms called microbes or pathogens that can multiply to larger numbers or produce large amounts of different types of toxins in food, which can then expose large numbers of persons across a wide range of food distribution. Viruses are also biological hazards because they infect their host and cause similar diseases, but they do not grow in food. Thus, they generally cause more limited geographic foodborne illness outbreaks related to working sick employees; the exception is when sick employees that handle RTE foods in food manufacturing facilities contaminate products (e.g., produce) that are purchased by multiple foodservice

© Springer Nature Switzerland AG 2020
H. King, *Food Safety Management Systems*, Food Microbiology and Food Safety, https://doi.org/10.1007/978-3-030-44735-9_2

locations. We refer to most biological-associated foodborne illnesses as diseases because the agent infects (or intoxicates) the host, leading to a disease (i.e., an infectious disease) that can oftentimes be contagious to other persons. A notable exception is an ingested toxin, which is not contagious.

Chemical hazards are more common—especially those classified as allergens, which induce allergic reactions to food, or other harmful chemicals such as pesticides or toxins found in fish. All of these can be introduced into food during food manufacturing via cross-contact with allergens in the facility (or may be intentionally present in food that is not labeled properly to provide avoidance messaging to susceptible consumers) or via cross-contact during food prep in a foodservice facility. These foodborne illnesses are not normally called diseases because they are not contagious/infectious; however, the outcomes can be just as harmful to the individual. Neither biological nor chemical hazards are easily observed or detected in foods.

Physical hazards are more easily observed in food because they are physical contaminants often introduced into food during food processing. These can include metal shavings from a mixer, plastic pieces from a conveyor belt, or bones left in what is supposed to be boneless chicken, for example, and are the most common cause of foodborne injury. Of course, these are non-biological and non-contagious/non-infectious; nevertheless, they can also lead to chemical illness if the physical hazard is also a heavy metal (e.g., lead).

The CDC measures and reports the numbers of foodborne illnesses that occur every year through national surveillance systems and via outbreak investigations. The majority of the confirmed foodborne illnesses reported to the CDC by individual states or investigated by the CDC are due to biological hazards/microbial pathogens (considered infectious foodborne diseases); the CDC investigates only those associated with multistate outbreaks (two or more cases of foodborne illness caused by the same agent and same food source across two or more states). This is because infectious foodborne disease outbreaks can cause illnesses in a large number of persons during a short time period. Also, the large number of cases of disease in an outbreak with the same genetically matched pathogen sometimes enables more statistically robust analyses to determine the linked food item/source (e.g., from a restaurant chain or food manufacturing facility) in common. However, there are also large numbers of outbreaks and sporadic cases of foodborne illness due to unknown causes, and/or many caused by chemical or other agents including infectious diseases not yet identified.

Therefore, the CDC categorizes its national surveillance of all foodborne illnesses into two major groups (Table 2.1):

1. Those illnesses (diseases) caused by known foodborne pathogens—31 pathogens known to cause foodborne illness (Table 2.2). Many of these pathogens are tracked by public health systems in each state that also track other diseases and outbreaks due to non-foodborne origin.
2. Those illnesses caused by unspecified agents—agents with insufficient data to estimate agent-specific burden; known agents not yet identified as causing foodborne illness; microbes, chemicals, or other substances known to be in food

**Table 2.1** Estimated annual number of domestically acquired foodborne illnesses, hospitalizations, and deaths due to 31 pathogens and the unspecified agents transmitted through food in the United States

| Foodborne agents | Estimated annual number of illnesses | Estimated annual number of hospitalizations % | Estimated annual number of deaths Number | % | Number | % |
|---|---|---|---|---|---|---|
| **31 known** | 9.4 million | 20 | 55,961 | 44 | 1351 | 44 |
| **Unspecified agents** | 38.4 million | 80 | 71,878 | 56 | 1686 | 56 |
| **Total** | 47.8 million | 100 | 127,839 | 100 | 3037 | 100 |

Source: CDC (2011)

**Table 2.2** 31 known pathogens that cause foodborne illnesses in the United States

| Bacteria | Parasites | Viruses |
|---|---|---|
| *Bacillus cereus* (foodborne), *Brucella* spp., *Campylobacter* spp., *Clostridium botulinum*, *Clostridium perfringens*, STEC O157, STEC non-O157, ETEC (foodborne), diarrheagenic *E. coli* other than STEC and ETEC, *Listeria monocytogenes*, *Mycobacterium bovis*, *Salmonella* spp., non-typhoidal *S. enterica* serotype Typhi, *Shigella* spp., *Staphylococcus aureus*, *Streptococcus* spp. group A, *Vibrio cholerae*, toxigenic *V. vulnificus*, *V. parahaemolyticus*, *Vibrio* spp., *Yersinia enterocolitica* | *Cryptosporidium* spp., *Cyclospora cayetanensis*, *Giardia intestinalis*, *Toxoplasma gondii*, *Trichinella* spp. | Astrovirus, Norovirus, Hepatitis A virus, Rotavirus, Sapovirus |

Source: CDC (2011)

whose ability to cause illness is unproven; and agents not yet identified. Because it is impossible to "track" what isn't yet identified, estimates for this group of agents describe health effects or symptoms that they are most likely to cause, such as acute gastroenteritis (e.g., stomach flu).

Out of the 31 foodborne pathogens (biological) known to cause foodborne illnesses (Table 2.2), 91% of all the estimated illnesses in the United States are caused by five pathogens (Table 2.3), of which *Norovirus* is the biggest culprit, causing over 50% annually. Interestingly, 80% of all the estimated annual foodborne illnesses in the United States are caused by unspecified agents, while only 20% are caused by known pathogens (Table 2.2). Thus, there is a much larger burden of foodborne illness in the United States from unknown causes, which clearly suggests that more work needs to be done in the identification of these hazards so foodservice businesses can develop controls for them now with the use of new diagnostic technologies that do not require culture of pathogens, more accurate incidence rates might be observed (Marder et al. 2018).

One of the most common statements I hear when I ask restaurant operators if they are having or have had any food safety issues in their restaurants is "we haven't had any outbreaks that I can remember so no, we don't have any food safety issues." Then, when I ask them if they have had any of these issues:

**Table 2.3** Top five pathogens contributing to domestically acquired foodborne illnesses in the United States

| Pathogen | Estimated number of illnesses | 90% credible interval |
|---|---|---|
| Norovirus | 5,461,731 | 3,227,078–8,309,480 |
| Salmonella (non-typhoidal) | 1,027,561 | 644,786–1,679,667 |
| Clostridium perfringens | 965,958 | 192,316–2,483,309 |
| Campylobacter spp. | 845,024 | 337,031–1,611,083 |
| Staphylococcus aureus | 241,148 | 72,341–529,417 |

Source: CDC (2011)

- Customer complaints of illness, allergy, or injury from physical hazards in the last month
- Social media post/reports of illness from several customers in the last month
- Health department inspection initiated due to reports of customer illnesses to them
- Low grade/score on the most current health inspection report

    - Any critical violations called foodborne illness risk factors (this is the strongest "bellwether" of the risk)

they often reluctantly acknowledge that they might actually have had a food safety issue in the past. Many restaurant operators don't associate day-to-day common food safety compliance issues (risk factors; see below) in their operations with a customer's foodborne illness risk unless they experience a foodborne illness investigation—or afterward, when they have already caused a foodborne disease outbreak. However, based on numerous studies by the FDA, CDC, and other experts, a restaurant is more likely to cause a single sporadic case of foodborne illness in a single location than to experience a foodborne disease outbreak.

When I am asked to speak to a group (usually a food industry trade conference of foodservice operators or businesses) about how to manage food safety risk in foodservice, I often use this example to show the difference between foodborne illness (which is personal) and foodborne disease outbreak (which is public). I ask the audience (in this example, let's say there are 100 attendees in the audience) to raise their hands if they have every experienced a foodborne illness AND they can guarantee they know what restaurant they got food poisoning from. Regularly, every time I do this, all attendees raise their hands. Then I ask the attendees to raise their hands again if they were part of a known foodborne disease outbreak investigated by a state health department or the CDC. Again, regularly no attendee raises their hands; suggesting that there are more sporadic cases of foodborne illnesses than there are cases from foodborne disease outbreaks.

In support of these observations, the Interagency Food Safety Analytics Collaboration (IFSAC, a forum of CDC, FDA, and USDA experts) has shown that there can be significantly more sporadic cases of foodborne disease (single-person, non-outbreak-associated illness) caused by the bacterial pathogens *Campylobacter*, *E. coli* O157, *Listeria monocytogenes*, and *Salmonella* each year than outbreak-associated cases caused by these same pathogens (Ebel et al. 2016). A study of foodborne

**Table 2.4** Differences between number of illnesses in all outbreaks vs. number of illnesses in all sporadic foodborne disease incidents in the United States between 2004 and 2011

| Pathogen | Outbreak illnesses | Sporadic illnesses | Outbreak fraction[a] |
|---|---|---|---|
| Campylobacter | 195 | 42,744 | 0.5% |
| E. coli O157 | 730 | 3117 | 19.0% |
| Listeria monocytogenes | 56 | 1024 | 5.2% |
| Salmonella | 3161 | 50,690 | 5.9% |

Data reproduced from Ebel et al. (2016)

[a]Outbreak fraction equals percent of all illnesses in the United States only associated with an outbreak

illnesses in the United States, analyzed by the IFSAC using FoodNet data (representing only 15% of the US population), identified 195 cases of *Campylobacter* illness associated with outbreaks between 2004 and 2011; by contrast, there were 42,744 single, sporadic cases of the same illness during this same time period (Table 2.4). Through extrapolation, one could speculate that there were likely 10 times these numbers of sporadic cases in the United States if 100% of the United States population had been represented in this surveillance. This strongly suggests that the total number of foodborne illnesses associated with sporadic events in the United States far outnumbers those from outbreaks.

A more direct example of sporadic vs. outbreak illness differences in restaurants from the state of Minnesota showed that there were more sporadic cases of foodborne illness than outbreak cases during the same time period. During 2010–2015, 154 sporadic *Salmonella* illnesses (each one linked to one of 154 individual fast-food and full-service restaurants) were reported by the state and local health department from just two cities: Bloomington and Richfield, MN (none were associated with a foodborne disease outbreak, Appling et al. 2018). During this same time period (using data from the CDC NORS Dashboard), there were 14 outbreaks caused by *Salmonella* in fast-food and full-service restaurants, with only 119 illnesses in the whole state of Minnesota reported to the CDC. Interestingly, all of the 154 restaurants in Bloomington and Richfield, MN, that caused a sporadic case of *Salmonella* had been frequently identified as being out-of-compliance (by the local health department) in regard to two contributing factors (see Appendix A) associated with the cause of foodborne illnesses in restaurants: (1) keeping food contact surfaces cleaned and sanitized and (2) keeping hand washing sinks accessible and stocked with soap and towels.

More evidence that sporadic cases of foodborne illness are more significant than outbreaks comes from the large volume of customer complaints of foodborne illness on social media sites such as Twitter, Yelp, Facebook, Google, or the more directed crowdsourced social media app, iwaspoisoned.com. Daily complaints on these social media sites indicate that sporadic cases are a daily occurrence across the United States. Many customer complaints of food poisoning on social media sites are also confirmed by local public health officials, showing customer complaints on social media and crowdsourced apps about specific foodservice locations can be factual. For example I recall tracking the email reports available through free subscription from iwaspoisoned.com (based on restaurants near your location, I chose the Atlanta area). During a 1-week period in December 2017, I began receiving multiple emails describing each customer's illness complaint (all were related to diarrhea and/or vomiting that had occurred over the last few days)—each from one foodservice establishment in a well-known university. Just a day or two after I started getting the emails, the Georgia Department of Public Health announced that there was a Norovirus outbreak in this same foodservice establishment (for more details, see FoodSafetyNews(2017)athttps://www.foodsafetynews.com/2017/10/iwaspoisoned-com-alerts-officials-to-outbreak-at-georgia-tech/).

Some US cities monitor these data sources or have developed their own apps for this purpose to prioritize health inspections based on number of customer complaints of food poisoning. The city of Chicago's Department of Public Health provides a Twitter account called Foodborne Chicago (see https://twitter.com/foodbornechi) for consumers to make complaints about illness related to specific restaurants. The department then uses the information to target foodservice establishments where health inspections should be prioritized. In a more compelling study using a machine learning model for real-time detection of potential foodborne illnesses (based on anonymous and aggregated web search and location data of persons searching for terms related to food poisoning), the model was able to identify restaurant locations that were 3.1 times more likely to be associated with health inspection violations related to the cause of foodborne illnesses (Sadilek et al. 2018). This model showed that many of the customer complaints of foodborne illness were actual sporadic cases and not associated with a foodborne disease outbreak.

So—if your foodservice/restaurant business has had more than one customer complaint of illness during any time period or, more particularly, complaints related to the same time period (e.g., over 3 days) and same food product (e.g., chicken salad), you might be causing sporadic foodborne illnesses which have simply not been discovered, yet. Likewise, if any of these complaints of illness coincide with the presence of any one of the foodborne illness risk factors described below, it is almost certain that you are causing foodborne illnesses in your establishment even though you haven't experienced a foodborne disease outbreak.

## Foodborne Illness Risk and the Contributing Factors Associated with Foodservice Establishments

The best means for determining whether a restaurant is at *risk* of causing a foodborne illness (or is likely already causing foodborne illnesses, as described above) is to look for the presence of the *risk factors* and their contributing factors (also called root causes; see below) that lead to the majority of all foodborne illnesses in foodservice establishments. The CDC and the FDA have stated for many years that the primary risk factors associated with most foodborne diseases (primarily due to biological hazards) caused by foodservice establishments are due to:

1. **Food from unsafe sources**—e.g., produce is sourced from a farm without food safety controls in place to prevent contamination with such pathogens as *E. coli* O157.
2. **Poor personal hygiene**—e.g., an asymptomatic employee working with bare, unwashed hands contaminates food with *Norovirus*.
3. **Inadequate cooking**—e.g., employee does not cook chicken to 165 °F to kill pathogens (e.g., *Salmonella*) that can be present on raw chicken.
4. **Improper holding/time and temperature**—e.g., cooked beans are held in the temperature danger zone (41°F to 135° F), enabling production of bacterial toxins (e.g., *Clostridium perfringens*).
5. **Contaminated equipment/protection from contamination**—e.g., food containers used to hold raw chicken (e.g., which can contain *Campylobacter jejuni*) are not properly cleaned and sanitized before being used to hold RTE foods.

These risk factors and their presence in a foodservice establishment are what a health department inspector looks for, in part, during a FDA risk-based inspection. Although they are not the only risk factors that lead to foodborne illnesses, foodborne illness outbreak data analysis has demonstrated that they are the most probable in restaurants. A health inspector will also measure the presence of good retail practices such as proper pest control and cleaning and sanitation processes that, when absent, may also contribute to foodborne illnesses.

In order to reduce/control these risk factors in a foodservice establishment, one must first understand the contributing factors that lead to these risks. The CDC reports that there are at least 32 contributing factors (Brown et al. 2017) that lead to foodborne disease outbreaks in restaurants (of course, sporadic cases would be attributable to the same factors) organized into three types of contributing factors: contamination (factors that introduce or otherwise permit the contamination of foods by pathogens, chemicals, or allergens), proliferation (factors that allow growth or amplification of the pathogens and/or their production of toxins in foods), and survival (factors that allow for survival or failure to destroy or inactivate pathogens/toxins in the foods). For the complete list of all 32 contributing factor definitions, see Appendix A and updates by the CDC here: https://www.cdc.gov/nceh/ehs/nears/cf-definitions.htm. In the most recent data analysis of 114 foodborne disease outbreaks in a cohort of 11 states/counties performed by the CDC National Environmental Assessment Reporting System (CDC 2015), 80% of the outbreaks were caused by a contamination contributing factor, while 28% and 17% were caused by proliferation and survival contributing factors, respectively (Fig. 2.1). The top five risk factors (described above, 1–5) have multiple contributing factors that introduce a hazard into food and also increase the risk of this when a foodservice establishment does not control them via a Process HACCP-based FSMS (which includes a Prerequisite Control Program for all of the 32 contributing factors).

It might seem intuitive that just understanding how to prevent the 32 contributing factors would be sufficient to prevent all foodborne illnesses. However, understanding

**Fig. 2.1** Percent of each contributing factor category in 2015 from a total of 114 outbreaks reported to CDC NEARS. The data included in this 2015 summary were collected by NEARS participants in California; Connecticut; Davis County, Utah; Fairfax County, Virginia; Harris County, Texas; Minnesota; New York City; New York State; Rhode Island; Tennessee; and Wisconsin. (Source: CDC 2015)

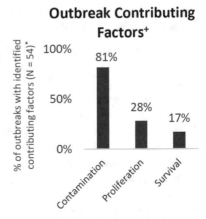

not only how each of the 32 contributing factors leads to one or more of the top five risk factors but also understanding where and when to implement the proper control is key to preventing the event from causing a foodborne illness. For example, there are multiple controls necessary to prevent many of the same contributing factors that lead contamination of food:

A. **Contributing factor**: Bare-hand contact with food by a food worker who has a foodborne illness and does not wash hands or wear gloves when preparing food **(Risk Factor Number 2)**
B. **Contributing factor**: Glove-hand contact with foods by a food worker who touches other surfaces bearing raw animal proteins while wearing gloves and then contaminates food during food prep **(Risk Factor Number 2)**
C. **Contributing factor**: Storage of food in a contaminated environment or container **(Risk Factor Number 5)**
D. **Contributing factor**: Food contaminated from a polluted source, for example, produce during harvesting **(Risk Factor Number 1)**

Clearly, the controls for each of these contributing factors would be different and necessary (and form part of the Process HACCP plan and Prerequisite Control Program; see Chap.3) in order to prevent the actual contamination event and would include, respectively:

A. **Control**: Restrict sick employees from working; ensure all employees wash their hands properly before donning gloves; ensure all employees wear gloves and/or use utensils when preparing RTE foods; clean and sanitize high-touch point surfaces especially the restroom door handlers, faucet knobs, etc. (see Chap. 4).
B. **Control**: Ensure all non-food contact surfaces are cleaned and sanitized, especially in kitchen prep areas where raw animal proteins are prepared; ensure

employees do not handle RTE foods unless they wash their hands and don new single-use gloves; use separate cleaning towels (or single-use wipes) stored in separate (color-coded) wet towel pails with sanitizer.

C. **Control**: Ensure all food storage containers are properly cleaned and sanitized; ensure all RTE foods are stored properly to ensure that no raw animal proteins drip/contact the foods.

D. **Control**: Source produce from farms/processors that are certified and audited to ensure prevention of contaminated fields/harvesters; check produce to confirm it is from approved source and at proper temperature upon receiving.

These are only a few of the 32 contributing factors (see Appendix A) that each relate to one or more of the top five risk factors (see above), of which each has led to a foodborne disease illness and/or outbreak from a restaurant each and every year that the CDC has studied outbreak causes. Thus, if foodservice establishments were to implement control of all of the contributing factors, they would not experience any of the top five risk factors, reducing the risk of causing a foodborne disease illness and/or an outbreak to the lowest feasible level. However, if a foodservice establishment has any one of these contributing factors during its day-to-day operations, it is likely contributing to foodborne illnesses—even if the establishment has not yet been associated with a foodborne disease outbreak. Finally, the good news and the purpose of this book is to demonstrate how easily these contributing factors can be controlled by foodservice establishments, thereby preventing foodborne illnesses and outbreaks.

# References

Appling XS, Lee P, Hedberg CW (2018) Risk factor violations associated with sporadic *Salmonella* cases. Front Public Health 6:355

Brown LG, Hover ER, Selman CA, Coleman EW, Rogers HS (2017) Outbreak characteristics associated with identification of contributing factors to foodborne illness outbreaks. Epidemiol Infect 145:2254–2262

CDC (2011) Estimates of foodborne illness in the United States. Content source: National Center for Emerging and Zoonotic Infectious Diseases (NCEZID), Division of Foodborne, Waterborne, and Environmental Diseases (DFWED).https://www.cdc.gov/foodborneburden/2011-foodborne-estimates.html

CDC (2015) CDC's National Environmental Assessment Reporting System (NEARS) 2015 summary report. https://www.cdc.gov/nceh/ehs/nears/docs/2015-nears-report.pdf

Ebel ED, Williams MS, Cole D, Travis CC, Klontz KC, Golden NJ, Hoekstra RM (2016) Comparing characteristics of sporadic and outbreak-associated foodborne illness, United States, 2004–2011. Emerg Infect Dis 22(7):1193–1200

Marder EP et al (2018) Preliminary incidence and trends of infections with pathogens transmitted commonly through food – foodborne diseases active surveillance network, 10 U. S. Sites, 2006-2017. Morb Mortal Wkly Rep 67:324–328

Sadilek A, Caty S, DiPrete L, Mansour R, Schenk T Jr, Bergtholdt M, Jha A, Ramaswami P, Gabrilovich E (2018) Machine-learned epidemiology: real-time detection of foodborne illness at scale. NPJ Digit Med 36:1–7

# Chapter 3
# The Process HACCP Plan and Prerequisite Control Program Necessary to Develop Food Safety Management Systems in Foodservice Establishments

Checking temperatures using paper checklist as the primary means to prevent food-borne illnesses and disease outbreaks does not prevent all known foodborne illness risk. In fact, there are more foodborne illnesses caused by employees working sick than those caused by uncooked or improperly held foods. This does not mean that ensuring foods are cooked, cooled, and held at the right temperatures is not preventive, but that it is not comprehensive to control all the hazards, and thus all the food safety risk in a foodservice establishment. The process approach to HACCP (called Process HACCP) recommended by the FDA is used to identify the hazards that occur during the food preparation processes of a complete menu and during actions made in the Prerequisite Control Program (e.g., employee personal hygiene, source of food ingredients, effective cleaning and sanitation SOPs, etc.). This is required in order to determine the most effective controls (CCPs and PCPs) for each of these hazards to prevent foodborne illnesses and outbreaks. The Food Safety Management Systems (FSMS) are then developed based on this Process HACCP plan and the Prerequisite Control Program, and then used by a Certified Food Protection Manager (CFPM) to monitor the controls during food preparation and other restaurant processes in the Prerequisite Control Program (described in detail in Chap. 4). The actual management of the hazards in foodservice operations via the monitoring, corrective actions, verifications, documentation, and validation functions of the FSMS can enable the foodservice business to establish Active Managerial Control of the risk factors that lead to most foodborne illnesses, injury (i.e., physical contaminants), and disease outbreaks.

## HACCP Is the Foundation of Prevention

Hazard Analysis and Critical Control Point (HACCP) is the most well-known and effective food safety management method in the food industry. HACCP principles have as their precursor an engineering system called Failure Mode and Effects

© Springer Nature Switzerland AG 2020
H. King, *Food Safety Management Systems*, Food Microbiology and Food Safety, https://doi.org/10.1007/978-3-030-44735-9_3

Analysis (FMEA) in the manufacture and validation of missiles by the US Army Laboratories at Natick MA (Mortimore and Wallace 2013). This system was used to look at what could potentially go wrong at each stage in an operation and the possible causes and probable effect of a failure. Then effective control mechanisms were put into place to ensure that the potential failures are prevented from occurring. In the early days of the US space program in the 1950s, HACCP was developed as a microbiological safety system to ensure the safety of food for astronauts where a foodborne illness could cause failure of a mission not to mention a horrible experience for an astronaut trapped in a space ship thousands of miles from Earth. At the time, most food safety systems were based on end-product testing which could only ensure safety if 100% of the end-product was tested—not feasible as no product would be left to sell.

The first HACCP system was developed to prevent hazards from getting into foods for astronauts by a team of food scientists and engineers from the Pillsbury Company, the US Army Laboratories at Natick, and the National Aeronautics and Space Administration (NASA). In 1971, Pillsbury presented this concept at the National Conference for Food Protection sponsored jointly by the FDA and the American Public Health Association. In 1974, the FDA incorporated the concepts of HACCP into its low-acid and acidified food regulations after several outbreaks of *Clostridium botulinum* (a lethal foodborne pathogen) from commercially canned foods. This first regulatory act in the United States to require HACCP in food manufacturing has virtually made canned foods safe for billions of people around the world, and whenever there is an outbreak of botulism from canned foods, it is because the business did not fully comply to the preventive HACCP program requirements (Surak 2009).

At the end of the 1980s, the National Research Council of the National Academy of Sciences published the recommendation (called *An Evaluation of the Role of Microbiological Criteria for Foods and Food Ingredients*) that all food processors and government regulatory authorities should use HACCP as the basis to ensure food safety in the US food supply. The publication also recommended that HACCP should be a regulatory requirement to ensure widespread use to prevent foodborne disease outbreaks in the United States, a recommendation that was not fully adopted until the Food Safety Modernization Act (FSMA) rules were enacted and became enforceable in 2017. During the 1980s and 1990s, HACCP use (voluntary) was driven by the marketplace as good for business by several companies like McDonald's that required all of their suppliers to implement HACCP systems; of which many other food manufacturers began to follow. This was one of the first examples where a large retail foodservice business began to influence its suppliers to improve food safety of its source of foods via requiring food safety specifications.

In November 1992, the National Advisory Committee on Microbiological Criteria for Foods (NACMCF) defined seven widely accepted HACCP principles that formally documented how a HACCP process should be implemented in a food manufacturing environment. In 1993 the Codex Alimentarius Commission issued its first international HACCP standard which provided the first definition of HACCP in the Codex Alimentarius. The Codex Alimentarius or "an international Food Code" is a collection of standards, guidelines, and codes of practice adopted by the Codex Alimentarius Commission. The commission, also known as CAC, is the central part of the Joint Food and Agriculture Organization of the United Nations/World Health Organization (FAO/WHO) Food Standards Program, and was established by FAO and WHO to protect consumer health and promote fair practices in international food trade. During this time, HACCP systems adoption changed many of the third-party auditing systems in use, and third-party and regulatory audits for HACCP compliance became the standard for good manufacturing practices (GMPs) in the food industry. In 1997, NACMCF reconvened to review the 1992 document and compare it to the then current HACCP guidance prepared by the Codex Committee on Food Hygiene. Based on this review, NACMCF endorsed HACCP use as its standard and defined HACCP as the best systematic approach to the identification, evaluation, and control of food safety hazards.

Today, HACCP has been implemented across the globe into many government and industry food safety standards such as the International Organization for Standardization (ISO) 22000 and the Global Food Safety Initiative (GFSI) which describe international state-of-the-art methods for application of HACCP for any part of the food supply chain developed by the world's food safety experts. Based on a solid foundation of Prerequisite Control Programs to control basic operational and sanitation conditions, the following seven basic principles of HACCP are now used to accomplish this objective:

Principle 1: Conduct a hazard analysis.
Principle 2: Determine the Critical Control Points (CCPs).
Principle 3: Establish critical limits.
Principle 4: Establish monitoring procedures.
Principle 5: Establish corrective actions.
Principle 6: Establish verification procedures.
Principle 7: Establish record-keeping and documentation procedures.

## HACCP in Retail Foodservice

HACCP use in restaurants in the United States has its roots in the US FDA Food Code. The FDA Food Code requires a comprehensive HACCP plan when conducting certain specialized food preparation processes for which production would otherwise be prohibited such as when a variance to a food preparation SOP is required by a health department and include, but are not limited to:

- Rendering of Time/Temperature Control for Safety (TCS) foods shelf stable
- Reduced oxygen packaging of TCS foods
- Curing of foods
- Smoking foods to render TCS foods shelf stable
- Use of molluscan shellfish tanks

- Sprouting of beans and seeds
- Slaughtering seafood at retail

The requirements for an HACCP plan for foodservice establishments operating under a variance to the current rules also have their foundation in the FDA Food Code. Most states use this as the standard to review and approve a variance to local and state food safety requirements. These HACCP plans (when needed by a foodservice business) must include flow diagrams, product formulations, training plans, and a corrective action plan based on the seven principles of HACCP described above (e.g., see Maricopa County's variance requirements using HACCP: https://www.maricopa.gov/3978/Food-Variances). These HACCP plans are then provided to the regulatory authority to enable the regulatory authority to assess whether the foodservice establishment has designed a system of controls sufficient to ensure the safety of the product. The plans will be reviewed, in most cases, in the absence of any microbiological performance information for the product at that establishment but can sometimes include requirement of the data to support efficacy of the controls.

HACCP is also the foundation of most health department inspections across the United States. The principles of HACCP are an integral part of the *FDA's Recommended Voluntary National Retail Food Regulatory Program Standards*. For regulatory program managers, the use of risk-based inspection methodology based on HACCP principles is a viable and practical option for evaluating the degree of Active Managerial Control operators have over the foodborne illness risk factors. The complete set of *Program Standards* is available from FDA through the following website:

http://www.fda.gov/Food/GuidanceRegulation/RetailFoodProtection/ProgramStandards/default.htm

Generally, however, the implementation of HACCP in restaurants is voluntary if no specialized processes of food preparation are used except in a few jurisdictions in some states and in the state of Maryland. For example, in Baltimore, MD, rules to operate a foodservice establishment regulated as a "high"- or "moderate"-risk establishment state that a HACCP plan must be submitted for each menu item sold by the foodservice establishment and must include:

1. Identification of Critical Control Points (CCPs) in each recipe/menu item. CCPs generally include cooking, cooling, reheating, cold holding, and hot holding, but other steps may be included if needed for a specific food. Hazards are controlled during those processes by following good retail practices (GRPs), sometimes referred to as standard operating procedures (SOPs).
2. Critical limits for each CCP.
3. Monitoring procedures for each CCP.
4. The corrective action that will be taken if there is a loss of control at a CCP due to such factors as employee error, equipment malfunction, or power failure.
5. Verification procedures that will ensure proper monitoring of each CCP, such as calibration of cooking and holding equipment and thermometers, and maintenance and review of records, such as temperature logs. Using logs for record-keeping is strongly encouraged, but not required, as long as the facility can

demonstrate that temperatures are routinely monitored, as described in the HACCP plan, and that specified corrective actions are taken when critical limits are not met.

6. A list of equipment used to support the proposed foodservice systems and maintain control at each CCP.
7. Written procedures for employee training on HACCP procedures.
8. The HACCP plan for a facility should be developed in a format which is easy for the employees to use. Once approved, this document must be readily available in the food preparation area of each facility. Examples of acceptable methods include:

   • Listing each CCP separately, with the menu items that utilize the CCP, critical limits, monitoring procedures, corrective action, verification methods for that CCP, and equipment used to control the CCP
   • Using an HACCP flow diagram and chart for selected menu items or groups of menu items
   • Incorporating each CCP and the monitoring procedures, corrective actions, and equipment used, directly into the recipe or preparation instructions used by the restaurant

These Baltimore, MD State health department requirements relate to each menu item/recipe so a restaurant business that has, for example, 50 food menu items (recipes) could be required to submit 50 HACCP plans and maintain these when any recipe changes and any new menu items are added. This use of HACCP is not incorrect as the identification of hazards and their controls is necessary. However, it is more appropriate for food manufacturing as it does not address all the hazards in foodservice that may be present in and outside of the food preparation processes in one plan.

As discussed in Chap. 1, applying HACCP to foodservice operations is very different than within a food manufacturing environment because of the high degree of variability and space constraints within this environment. The process approach to HACCP in foodservice environments (called Process HACCP) does not require a HACCP plan for each recipe/menu item as is used in the food manufacturing environment (FDA 2006). Process HACCP method enables the foodservice establishment to identify each of the hazards (the hazard analysis) at each process stage (from receiving, storing, preparing, cooking, reheating, hot holding, and serving foods) during the preparation of all menu items and during any process that occurs in the Prerequisite Control Program in to one plan (Fig. 3.1). All of the critical limits that must be controlled at each process are then defined to include each and every hazard including biological, chemical (allergens), and physical ones at each process. This achieves the same control of foodborne risk factors as if preparing a complete HACCP plan for each individual recipe/menu item. Because all food preparation processes may be used in a foodservice business at some point in the life of the business, the flow of food must be included even if a process, e.g., cooling down foods, does not occur.

The primary reason why the Process HACCP methods is so effective is because it uses a strong foundation of HACCP in the food preparation processes of a specific

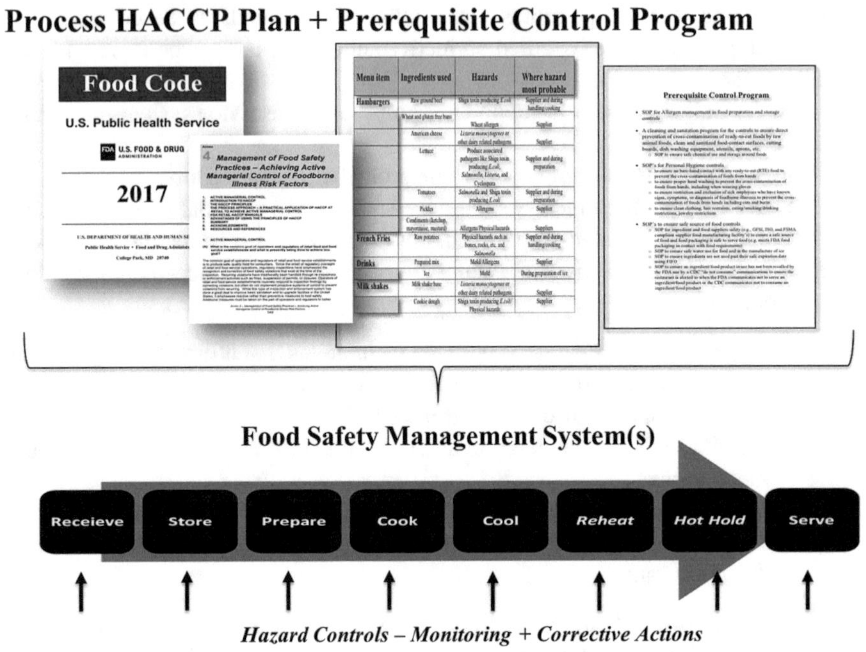

## Process HACCP Plan + Prerequisite Control Program

**Fig. 3.1** How the Process HACCP plan plus the Prerequisite Control Program (this chapter) are used to create Food Safety Management Systems (FSMS, Chap. 4). Once the hazards have been identified in the food preparation processes used to serve the menu and in the Prerequisite Control Program, and the controls have been identifed that must be in place to prevent the hazards, the FSMS are developed to enable the foodservice business to monitor the controls and take appropriate action to ensure each is in place

menu, and it includes a requirement to address additional hazards that occur during basic operational procedures such as how to ensure safe source of ingredients and products, employee food safety training, sanitation controls, a health policy to reduce risk of working sick employees, etc. within an operation. These procedures are collectively termed the "Prerequisite Control Program." When a Prerequisite Control Program is in place along with the Process HACCP plan, a more comprehensive means for controlling hazards associated with a foodservice business using FSMS is achieved.

As discussed in Chaps. 1 and 4, the Prerequisite Control Points (PCPs are used to differentiate where in and outside the food preparation processes controls must be monitored to prevent hazards) are derived from a predefined Prerequisite Control Program, and are primarily designed to place controls over source of food ingredients, employee personal hygiene, pest control, cross-contamination risk critical to the safe food preparation processes, etc. When a Prerequisite Control Program is defined and executed with the Process HACCP plan, the FSMS are more effective and more comprehensive in controlling all of the hazards associated with the menu a foodservice business prepares and sells. While the Process HACCP plan is developed to identify hazards (via the hazard analysis) and their controls, the Prerequisite Control Program is primarily designed to identify known controls of contributing factors necessary to ensure the safe preparation of food. The minimum recommended components of a Prerequisite Control Program that will provide more comprehensive controls in FSMS (see Chap. 4) include:

- Methods to ensure equipment maintenance

  - Equipment to ensure temperature measuring devices are calibrated.
  - Cooking equipment is calibrated, and hot and cold holding equipment provides the correct temperature.
  - Refrigeration and freezer equipment provide the correct environmental temperature.
  - Ware washing equipment are operating according to manufacturer's specifications.

- Methods to ensure allergen management in food preparation and storage
- Methods to ensure safe chemical use and storage around foods and employee safety that meet Occupational Safety and Health Administration (OSHA) requirements for foodservice businesses (see https://www.osha.gov/SLTC/restaurant/)
- Methods to ensure safe water use for food and in the manufacture of ice
- An effective pest prevention program to prevent pest infestations
- Methods to ensure no bare-hand contact with any ready-to-eat (RTE) food to prevent the cross-contamination of foods from hands
- Methods to ensure proper hand washing to prevent the cross-contamination of foods from hands, including when wearing gloves
- Methods to ensure restriction and exclusion of sick employees who have known signs, symptoms, or diagnosis of foodborne illnesses to prevent the cross-contamination of foods from hands, including cuts and burns (as part of the health policy; see Chap. 4)
- Personal hygiene requirements of employees (clean clothing, hair restraints, eating/smoking/drinking restrictions, jewelry restrictions)
- A cleaning and sanitation program for the direct prevention of cross-contamination of ready-to-eat foods by raw animal foods to clean and sanitized food contact surfaces, cutting boards, dish washing equipment, utensils, aprons, etc.
- Methods to ensure safe source of foods—using only food safe ingredient and food suppliers (e.g., GFSI-, ISO-, and FSMA-compliant supplier food manufacturing facilities) to ensure a safe source of food, and food packaging that is safe to serve food on (e.g., meets FDA food packaging in contact with food requirements)
- Methods to ensure ingredients are not used past their safe expiration date using FIFO
- Methods to ensure an ingredient/food product in use has not been recalled by the FDA nor by a CDC "do not consume" communications to ensure the restaurant

is alerted to when the FDA communicates not to serve an ingredient/food product or the CDC communicates not to consume an ingredient/food product

These last three bullets in a Prerequisite Control Program manage one of the top five risk factors, (unsafe ingredient and food sources) that when not under control can increase the number of hazards in all other processes (and has led to many foodborne disease outbreaks from retail foodservice establishments). For example, if flour (considered a raw agricultural product by the FDA) is to be used in food preparation processes such as dusting food prep surfaces to keep RTE foods from sticking or for preparation of cookie dough, it must be sourced from a supplier that will provide heat-treated flour for this purpose. Otherwise, the use of raw flour in this manner can lead to a foodborne illness outbreak (e.g., see CDC 2019) because one of the hazards associated with raw flour is *E. coli* O157 and *Salmonella* contamination. Of course if the raw flour will only be used for preparing baked goods, it does not necessarily need to be heat-treated by a supplier. Many multi-location foodservice businesses have established supply chain food safety management and procurement programs to ensure that food ingredients and products used by their establishments (based on how they are used to prepare menu items) are safe according to national and international certification standards (e.g., GFSI- and/or ISO-certified food manufacturing facilities) and meet the brands' quality specifications. These businesses also audit (as part of food safety management) their high-risk supplier's facilities as well.

Many small businesses (e.g., growing franchised restaurant chain business) may not be sufficiently established or have the means to build a supply chain food safety management system to ensure a complete Prerequisite Control Program that will ensure all the proper controls of expected hazards are in place and managed by their suppliers. Likewise, many independent businesses (e.g., new restaurant concept) do not have the experience in how to source safe food ingredients and products via certification of suppliers nor auditing of their suppliers, and thus they rely on broadline or local food distributors to source all of their food. However there are other means to ensure safe source of foods for small and single-location foodservice businesses using a process required by the FDA of all human food manufacturing facilities in the United States. Recent enhancements to FDA regulation of manufactured human foods under the Food Safety Modernization Act (FSMA) are improving the means for independent and/or small chain foodservice establishments, and even corporate businesses without a well-established supplier food safety management program, to ensure safe food sources from their suppliers. The act mandates that food manufacturers adhere to a food safety management system similar to Process HACCP to achieve Active Managerial Control of risk called Hazard Analysis and Risk-Based Preventive Controls (HARPC; see King and Ades (2015)). First, the foodservice business can ask the supplier for a copy of the FDA required Food Safety Plan for any ingredient and food product manufactured for or purchased by the business. The Food Safety Plan is a HARPC requirement where the supplier must list all the ingredients used to manufacture a food product, list all facility- and ingredient-related hazards associated with each ingredient and process, and then define what controls will be in place during the manufacture of that food product at each and every product run. The supplier must also keep record of the monitoring of each control inclusive of any corrective actions, and be able to document this during an FDA inspection of the facility. Second, because the FDA (or a state government regulatory partner) only inspects the food facility once every 3 years, a foodservice business can ask to see the most current FDA inspection results and the most recent supplier monitoring documentation related to the Food Safety Plan of any food product they purchase from a supplier as a means to verify safe source of foods. For more detailed explanations and procedures on how to use HARPC to ensure safe source of foods, see the book *Hazard Analysis and Risk-Based Preventive Controls: Improving Food Safety in Human Food Manufacturing for Food Businesses* (King and Bedale 2017).

Because the Prerequisite Control Program is critical to the safe preparation and service of food, the Food Safety Management System must include the establishment and the inclusion of the controls of each PCP, its critical limits, and a means to monitor the control as is also done with the Process HACCP controls at each process. For example, the FSMS should define when employees must wear single-use foodservice gloves when handling RTE foods, and provide a tool to monitor the proper use of gloves (see King and Michaels (2019) on a daily basis including what to do as a corrective action when managers observe an employee not using gloves properly to prevent cross-contamination of foods during the food prep processes. These PCPs and the proper controls within a food safety management system are discussed in more detail in Chap. 4. The hazards that can occur without determining PCPs (see below) and their controls, e.g., an employee working while sick with *Norovirus*, and not wearing single-use foodservice gloves when handling RTE foods, can lead to even larger foodborne disease outbreaks than a missed CCP control, such as not cooking a single chicken breast properly during the food prep process, which likely would only make one customer sick. Both controls are important of course, and this is why the Food Safety Management System must include PCPs specific to personal hygiene and other controls within the Prerequisite Control Program including cleaning and sanitation PCPs such as reducing transmission of Norovirus on high-touch point surfaces as discussed in Chap. 4.

## Hazards in Food Preparation Processes and Their Effective Controls Using CCPs and PCPs

As discussed in Chaps. 1 and 2, there are three primary types of hazards in the preparation of food in a foodservice establishment: biological, chemical, and physical hazards. Each of these hazards can be introduced into the food before it is received into the restaurant (e.g., at a food manufacturing facility or on a farm) and also while the food is being stored, prepared, and served when in the restaurant by employees' hands, cross-contamination, etc. Understanding what the hazards are and how and when they may contaminate foods you serve to customers is critical to the effectiveness of the FSMS. Interestingly, many of the same hazards like *Salmonella* can be introduced anywhere along the food chain including during foodservice and by an employee with a *Salmonella* infection that does not wear gloves properly (Fig. 3.2). In order to properly include all the known hazards in the

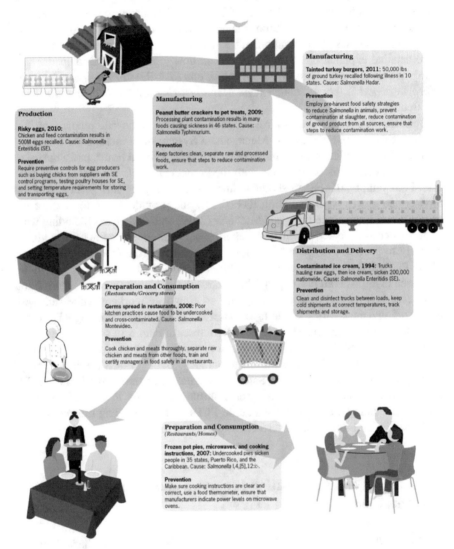

**Manufacturing**

**Tainted turkey burgers, 2011:** 50,000 lbs of ground turkey recalled following illness in 10 states. Cause: *Salmonella* Hadar.

**Prevention**
Employ pre-harvest food safety strategies to reduce *Salmonella* in animals, prevent contamination at slaughter, reduce contamination of ground product from all sources, ensure that steps to reduce contamination work.

**Manufacturing**

**Peanut butter crackers to pet treats, 2009:** Processing plant contamination results in many foods causing sickness in 46 states. Cause: *Salmonella* Typhimurium.

**Prevention**
Keep factories clean, separate raw and processed foods, ensure that steps to reduce contamination work.

**Production**

**Risky eggs, 2010:** Chicken and feed contamination results in 500M eggs recalled. Cause: *Salmonella* Enteritidis (SE).

**Prevention**
Require preventive controls for egg producers such as buying chicks from suppliers with SE control programs, testing poultry houses for SE, and setting temperature requirements for storing and transporting eggs.

**Distribution and Delivery**

**Contaminated ice cream, 1994:** Trucks hauling raw eggs, then ice cream, sicken 200,000 nationwide. Cause: *Salmonella* Enteritidis (SE).

**Prevention**
Clean and disinfect trucks between loads, keep cold shipments at correct temperatures, track shipments and storage.

**Preparation and Consumption**
*(Restaurants/Grocery stores)*

**Germs spread in restaurants, 2008:** Poor kitchen practices cause food to be undercooked and cross-contaminated. Cause: *Salmonella* Montevideo.

**Prevention**
Cook chicken and meats thoroughly, separate raw chicken and meats from other foods, train and certify managers in food safety in all restaurants.

**Preparation and Consumption**
*(Restaurants/Homes)*

**Frozen pot pies, microwaves, and cooking instructions, 2007:** Undercooked pies sicken people in 35 states, Puerto Rico, and the Caribbean. Cause: *Salmonella* I,4,[5],12:i:‑.

**Prevention**
Make sure cooking instructions are clear and correct, use a food thermometer, ensure that manufacturers indicate power levels on microwave ovens.

**Fig. 3.2** Understanding where a single hazard such as the bacterial pathogen *Salmonella* spp. (with multiple species and serotypes) can contaminate food along the supply chain including at foodservice. It is important to define where the hazard is most likely to be introduced to a food during the Hazard Analysis process (as part of the Process HACCP plan and Prerequisite Control Program development) to best enable the effective control of that hazard. The risk of a single hazard can be significantly reduced when there are multiple controls of that hazard throughout the supply chain, and during foodservice food preparation

Process HACCP plan, we first must list each and then define how they might be controlled as a PCP or CCP by a supplier and/or during daily operations in the food-service establishment. Remember that many of the same hazard types may contaminate food at different points in the supply chain and then in the restaurant by multiple contributing factors, and therefore must be controlled by both PCPs and CCPs. For instance, the biological hazard *Salmonella* Typhi is likely to be found on raw chicken parts (introduced during raw chicken processing by a supplier) which then can contaminate food in the restaurant, and would need to *be prevented by multiple controls* **and their respective corrective actions** (e.g., each foodservice business should determine the corrective actions specific to its food processes and menu) to fully prevent each. For example, these contributing factors require different controls and corrective actions:

- Thawed or thawing raw chicken dripping on RTE food during storage on the upper shelf in a walk-in refrigerator, and the RTE food is not cooked further.

  – *Control storing the raw chicken on the lowest shelf under all RTE foods*

    **Corrective action discard any RTE foods not covered or that raw chicken juice has dripped on and move the raw chicken to the lowest shelf, and note if employee requires re-training (if prior history of observed out-of-compliance).**

- The raw chicken is not thawed properly and then not cooked to 165 °F fully to kill *Salmonella* Typhi because of partial thawing.

  – *Control thawing the raw chicken properly using proper time and temperature (not above 41 °F)—if thawing and then cooking times are based on size/thickness of the chicken parts, a control for proper size/thickness as a specification for the raw chicken supplier is necessary.*

    **Corrective action check thaw of frozen chicken after thawing to ensure no frozen parts checking center of piece if boneless filet and around bones if not. Continue thawing at or below 41 °F until chicken is completely thawed before cooking.**
    **Check size and weight of chicken and compare to raw chicken supplier specification and notify supervisor/manager if not to specification.**

  – *Control cooking the chicken to 165 °F.*

    **Corrective action recook the chicken for 1–3 min and recheck the temperature a second time; if under 165 °F, discard chicken, stop use of cooking equipment, and notify supervisor/manager to determine root cause of undercooked chicken.**

- An employee handling the raw chicken without gloves during prep does not wash his/her hands and touches other surfaces in the kitchen that other employees also touch, who then handle RTE food.

- *Control employee wears gloves during handling of raw chicken, removes gloves, and washes hands after food prep before touching other surfaces in the kitchen.*

  **Corrective action observe employee while handling raw chicken and moving to other tasks, have employee clean and sanitize any surface touched by hands and rewash hands properly, and note if employee requires re-training (if prior history of observed out-of-compliance).**

- An employee cleans a surface used to prepare raw chicken with a reusable towel and then places the towel into a pail of sanitizer solution for storage that is also then used to clean RTE food contact surfaces.

  - *Control employee uses a designated reusable towel and storage pail (yellow for raw chicken) for cleaning up raw chicken prep surfaces that is not used for cleaning and sanitizing RTE food prep surfaces.*

    **Corrective action have employee discard reusable towel or place in dirty towel bin and replace with reusable towel stored in yellow pail of sanitizer for use on raw chicken prep surfaces only, and note if employee requires re-training (if prior history of observed out-of-compliance).**

- An employee does not properly clean and sanitize a food contact surface after preparing raw chicken, and then another employee prepares an RTE food on this same surface.

  - *Control the employee cleans and sanitizes the food contact surface after preparing raw chicken.*

    **Corrective action discard any prepared RTE food made on this surface (including if already stored), have employee clean and sanitize the surface, and note if employee requires re-training (if prior history of observed out-of-compliance).**

The best way to determine what hazards and their controls are likely to be found in a foodservice facility is to source this information from the most current FDA Food Code (2017 at the time of this writing), and augment this with information from recent foodborne disease outbreaks investigated by the CDC (see https://www.cdc.gov/foodsafety/outbreaks/index.html). For example, the known biological, chemical (which includes allergen hazards), and physical hazards and their control measures are listed in Tables 3.1, 3.2, and 3.3, respectively, where the additional hazards added from new knowledge up to early 2020 are in italics. Each control must be validated by science (FDA, CDC, and academic publication by experts) and be accepted by the FDA as an effective control. For instance, the cooking temperature of raw poultry has been determined to be 165 °F by experts that evaluated all the most common pathogens (e.g., *Salmonella, Listeria monocytogenes*, etc.) on raw poultry to ensure all the variables including time, numbers of bacteria, etc. have been validated. Note that although many of these hazards will not change, new ones may be discovered or new control methods validated by the FDA. For example, the biological parasitic hazard *Cyclospora* causing the illness cyclosporiasis is not dis-

**Table 3.1** Biological hazards, their associated ingredients/foods, and control measures in foodservice businesses

| Hazard | Associated foods | Control measures |
|---|---|---|
| **Bacteria** | | |
| *Bacillus cereus* (intoxication caused by heat-stable, preformed emetic toxin and infection by heat-labile, diarrheal toxin) | Meat, poultry, starchy foods (rice, potatoes), puddings, soups, cooked vegetables | Cooking, cooling, cold holding, hot holding |
| *Campylobacter jejuni* | Poultry, raw milk | Cooking, hand washing, prevention of cross-contamination |
| *Clostridium botulinum* | Vacuum-packed foods, reduced oxygen packaged foods, under-processed canned foods, garlic-in-oil mixtures, time/temperature abused baked potatoes/sautéed onions | Thermal processing (time + pressure), cooling, cold holding, hot holding, acidification and drying, etc. |
| *Clostridium perfringens* | Cooked meat and poultry, cooked meat and poultry products including casseroles, gravies | Cooling, cold holding, reheating, hot holding |
| *E. coli* O157:H7 (other Shiga-toxin-producing *E. coli*) | Raw ground beef, raw seed sprouts, raw milk, unpasteurized juice, foods contaminated by infected food workers via fecal-oral route | Cooking, no bare-hand contact with RTE foods, employee health policy, hand washing, prevention of cross-contamination, pasteurization or treatment of juice |
| *Listeria monocytogenes* | Raw meat and poultry, fresh soft cheese, paté, smoked seafood, deli meats, deli salads | Cooking, date marking, cold holding, hand washing, prevention of cross-contamination |
| *Salmonella* spp. | Meat and poultry, seafood, eggs, raw seed sprouts, raw vegetables, raw milk, unpasteurized juice | Cooking, use of pasteurized eggs, employee health policy, no bare-hand contact with RTE foods, hand washing, pasteurization or treatment of juice |
| *Shigella* spp. | Raw vegetables and herbs, other foods contaminated by infected workers via fecal-oral route | Cooking, no bare-hand contact with RTE foods, employee health policy, hand washing |
| *Staphylococcus aureus* (preformed heat-stable toxin) | RTE PHF foods touched by bare hands after cooking and further time/temperature abused | Cooling, cold holding, hot holding, no bare-hand contact with RTE food, hand washing |
| *Vibrio* spp. | Seafood, shellfish | Cooking, approved source, prevention of cross-contamination, cold holding |

(continued)

**Table 3.1** (continued)

| Hazard | Associated foods | Control measures |
|---|---|---|
| **Parasites** | | |
| *Anisakis simplex* | Various fish (cod, haddock, fluke, Pacific salmon, herring, flounder, monkfish) | Cooking, freezing |
| *Taenia* spp. | Beef and pork | Cooking |
| *Trichinella spiralis* | Pork, bear, and seal meat | Cooking |
| **Viruses** | | |
| Hepatitis A and E | Shellfish, any food contaminated by infected worker via fecal-oral route | Approved source, no bare-hand contact with RTE food, minimizing bare-hand contact with foods not RTE, employee health policy, hand washing |
| Other viruses (rotavirus, Norovirus, reovirus) | Any food contaminated by infected worker via fecal-oral route | No bare-hand contact with RTE food, minimizing bare-hand contact with foods not RTE, employee health policy, hand washing |

Source: Annex 4, Table 1. FDA Food Code, FDA 2017
*RTE* ready-to-eat, *PHF* potentially hazardous food, *TCS* Time/Temperature Control for Safety food

cussed as a significant hazard in produce in the 2017 FDA Food Code, but it has caused several outbreaks since 2018 due to contaminated produce (see https://www.cdc.gov/parasites/cyclosporiasis/outbreaks/index.html).

Other useful resources include the academic programs in colleges and universities like the Colorado State University Food Source Information website/Wiki (see https://fsi.colostate.edu). According to the website, the Food Source Information Wiki was developed by Colorado State University (CSU), in collaboration with the Colorado School of Public Health (CSPH) and the Colorado Department of Public Health and Environment (CDPHE), as part of the Colorado Integrated Food Safety Center of Excellence. The purpose of the Food Source Information Wiki is to provide public health professionals with rapid access to basic information on production practices and food distribution systems for a range of agricultural food products, from farm to fork. Basic and timely information on agricultural production practices for foods suspected or implicated during a foodborne illness outbreak has better equipped outbreak responders to determine the cause of the outbreak and contributing factors. This information is vital for preventing further illnesses and outbreaks but can also be used by foodservice businesses to research hazards associated with food ingredients like celery or potatoes to ensure a comprehensive hazard analysis in the Process HACCP plan. Likewise, several colleges and universities have food safety research programs that validate new hazards and their controls in the supply chain, and oftentimes publish new findings important to the management of risk to foodservice businesses. The International Association for Food Protection (see https://www.foodprotection.org) is an excellent resource to keep up with the latest information about new hazards and controls published in their scientific journals,

**Table 3.2** Chemical hazards, their associated ingredients/foods, and control measures in foodservice businesses

| Material | Injury potential | Sources |
|---|---|---|
| Scombrotoxin | Primarily associated with tuna fish, mahi-mahi, blue fish, anchovies bonito, mackerel; it is also found in cheese | Check temperatures at receiving; store at proper cold holding temperatures. Buyer specification: obtain verification from supplier that product has not been temperature abused prior to arrival in facility |
| Shellfish toxins | Molluscan shellfish from NE and | Ensure all molluscan shellfish are from approved sources and properly tagged and labeled according to the FDA seafood HACCP requirements |
| Paralytic shellfish poisoning (PSP) | NW coastal regions of the United States, viscera of lobsters, and Dungeness, tanner, and red rock crabs | |
| Diarrhetic shellfish poisoning (DSP) | Molluscan shellfish in Japan, Western Europe, Chile, Eastern Canada, | |
| Neurotoxin shellfish poisoning (NSP) | Gulf of Mexico | |
| Amnesic shellfish poisoning (ASP) | Molluscan shellfish from NE and NW coastal regions of the United States; viscera of Dungeness, tanner, and red rock crabs; and anchovies | |
| Ciguatoxin | Reef fin fish from extreme SE United States, Hawaii, and tropical areas, barracuda, jacks, king mackerel, large groupers, and snappers | Ensure fin fish have not been caught: • _Purchase fish from approved sources. • _Fish should not be harvested from an area that is subject to an adverse advisory |
| Tetrodotoxin | Puffer fish (fugu, blowfish) | Do not consume these fish |
| Mycotoxins aflatoxin and patulin | Corn and corn products, peanuts and peanut products, cottonseed, milk, and tree nuts such as Brazil nuts, pecans, pistachio nuts, and walnuts. Other grains and nuts are susceptible but less prone to contamination. Apple juice products | Check condition at receiving; do not use moldy or decomposed food. Buyer specification: obtain verification from supplier or avoid the use of rotten apples in juice manufacturing |
| Toxic mushroom species | Numerous varieties of wild mushrooms | Do not eat unknown varieties or mushrooms from unapproved source |

(continued)

**Table 3.2** (continued)

| Material | Injury potential | Sources |
|---|---|---|
| Pyrrolizidine alkaloids | Plant foods containing these alkaloids. Most commonly found in members of the Boraginaceae, Compositae, and Leguminosae families | Do not consume food or medicinals contaminated with these alkaloids |
| Phytohemagglutinin | Raw red kidney beans (undercooked beans may be more toxic than raw beans) | Soak in water for at least 5 h. Pour away the water. Boil briskly in fresh water, with occasional stirring, for at least 10 min |
| Environmental contaminants: Pesticides, fungicides, fertilizers, insecticides, antibiotics, growth hormones | Any food may become contaminated | Follow label instructions for use of environmental chemicals. Soil or water analysis may be used to verify safety |
| PCBs | Fish | Comply with fish advisories |
| Prohibited substances (21 CFR 189) | Numerous substances are prohibited from use in human food; no substance may be used in human food unless it meets all applicable requirements of the FD&C Act | Do not use chemical substances that are not approved for use in human food |
| Toxic elements/compounds of mercury | Fish exposed to organic mercury: shark, tilefish, king mackerel, and swordfish. Grains treated with mercury-based fungicides | Pregnant women/women of childbearing age/nursing mothers, and young children should not eat shark, swordfish, king mackerel, or tilefish because they contain high levels of mercury. Do not use mercury-containing fungicides on grains or animals |
| Copper | High-acid foods and beverages | Do not store high-acid foods in copper utensils; use backflow prevention device on beverage vending machines |
| Lead | High-acid food and beverages | Do not use vessels containing lead |
| Preservatives and food additives: sulfiting agents (sulfur dioxide, sodium and potassium bisulfite, sodium and potassium metabisulfite) | Fresh fruits and vegetables, shrimp, lobster, wine | Sulfiting agents added to a product in a processing plant must be declared on labeling. Do not use on raw produce in food establishments |

(continued)

**Table 3.2** (continued)

| Material | Injury potential | Sources |
|---|---|---|
| Nitrites/nitrates niacin | Cured meats, fish, any food exposed to accidental contamination, spinach<br><br>Meat and other foods to which sodium nicotinate is added | Do not use more than the prescribed amount of curing compound according to labeling instructions. Sodium nicotinate (niacin) is not currently approved for use in meat or poultry with or without nitrates or nitrates |
| Chemicals used in retail establishments (e.g., lubricants, cleaners, sanitizers, cleaning compounds, and paints | Any food could become contaminated | Address through SOPs for proper labeling, storage, handling, and use of chemicals; retain material safety data sheets for all chemicals |
| Chemicals used in foodservice establishments (lubricants, cleaners, sanitizers, cleaning/ scrubbing compounds, floor cleaners, disinfectants, pesticides, and paints) | See safety data sheets (SDS) for chemical safety and toxicity to humans | Ensure employee training and proper PPE are available as required by the SDS |

Source: Annex 4, Table 1. FDA Food Code, FDA (2017)

**Table 3.3** Physical hazards, their associated ingredients/foods, and control measures in foodservice businesses

| Material | Injury potential | Sources |
|---|---|---|
| Glass fixtures | Cuts, bleeding, may require surgery to find or remove | Bottles, jars, lights, utensils, gauge covers |
| Wood | Cuts, infection, choking, may require surgery to remove | Fields, pallets, boxes, buildings |
| Stones, metal fragments | Choking, broken teeth, cuts, infection, may require surgery to remove | Fields, buildings, machinery, wire, employees |
| Insulation | Choking, long-term if asbestos | Building materials |
| Bone | Choking, trauma | Fields, improper plant processing |
| Plastic | Choking, cuts, infection, may require surgery to remove | Fields, plant packaging materials, pallets, employees |
| Personal effects | Choking, cuts, broken teeth, may require surgery to remove | Employees |

Source: Annex 4, Table 1. FDA Food Code, FDA (2017)

and many science publishers also publish journals specific to food safety, including the publisher of this book. I recommend the food safety professional use this information and also follow the most current CDC and FDA investigations (posted on the CDC/FDA websites) on a regular basis as new hazards/food commodities/production methods and means of contamination may be discovered that should be defined in the Process HACCP plan.

In addition to the chemical hazards listed in Table 3.2, many ingredients have been reported to cause food allergies. The FDA believes there is scientific consensus that the following foods can cause a serious allergic reaction in sensitive individuals, as these foods account for 90% or more of all human food allergies:

- Milk
- Egg
- Fish (such as bass, flounder, or cod)
- Crustacean shellfish (such as crab, lobster, or shrimp)
- Tree nuts (such as almonds, pecans, or walnuts)
- Wheat
- Peanuts
- Soybeans
- *Sesame* (not currently in the FDA list but should be; see below)

As discussed above, the food safety professional must continually review the scientific literature and CDC/FDA reports and investigations throughout the year to ensure new hazards are not missed that should be included in the Process HACCP plan or as part of the Prerequisite Control Program. For example, additional allergens such as sesame are likely important to add as a chemical hazard even though it is not listed by the FDA currently as a significant hazard. As an ingredient, sesame seeds are used widely in foodservice—they can be found in tahini and sushi or often mixed in granola, sprinkled on bagels and other breads, or used as a flavoring in several dishes. While previous studies have suggested sesame allergies affected about 0.2% of US children and adults, new research published in the *Journal of the American Medical Association* (Warren et al. 2019) estimated that the number of sesame-allergic Americans could be as high as 0.49% which represents around 1.6 million people. The study's findings also were published at a time when the FDA was considering adding sesame to its list of top allergens that must be noted on food packaging, issuing a request for information on the "prevalence and severity" of sesame allergies in the United States to aid in its decision. Thus, even though sesame allergens are not regulated (e.g., via labeling requirements) as yet, a foodservice business must determine if a hazard like this allergen should be defined in its Process HACCP plan or Prerequisite Control Program based on the probability of it being in the foods they serve (i.e., ingredients they source including if "manufactured in a facility with" the allergy) and the likelihood a customer may be exposed to the allergen without avoidance messaging provided (King and Bedale 2018). In this scenario, I would recommend sesame allergen be added to the hazard analysis for your plan.

## Design of the Process HACCP Plan and Prerequisite Control Program

The steps in the design of the Process HACCP plan and the Prerequisite Control Program and then their use in development of the FSMS (Chap. 4) begin with grouping all the menu items and their recipes prepared and served to customers into the processes used in the restaurant. Because most restaurants perform five or more of the common food preparation processes shown in Fig. 3.1 (which includes all three food preparation process types described by the FDA), each menu item recipe can be categorized into each of these food preparation processes to document where a hazard may occur and where and how to place the control(s) regardless of what type of restaurant concept it is. Note that the hazard analysis must also be performed for the Prerequisite Control Program so that controls can be defined for such hazards related to source of food ingredients (which is a part of the program for selecting food ingredient/product suppliers, evaluating their preventive controls, receiving foods, FIFO, and date of expiration management), employee personal hygiene, etc. (see Prerequisite Control Program descriptions above) and can be used in the FSMS and monitored at the appropriate times during operations. It is not in the scope of this book to define all the components of a Prerequisite Control Program and the hazards as there are many variables related to each based on different suppliers, and menu type and specific to the food preparation processes (as PCPs common to all foodservice businesses). The development of Prerequisite Control Programs is discussed at length in my first book (King 2013).

Using the menu items being served in a foodservice business, the steps used to establish the Process HACCP plan are as follows:

1. Use all menu items and prepare a list of ingredients from the recipes used to prepare each and the source of each ingredient. The ingredients, their type and source, and the processes used to prepare ingredients into a menu item will inform the most likely hazards.
2. List the hazards associated with the preparation of each recipe and where the hazard is most probable.
3. List the primary control of each hazard and how it will be executed (either as a CCP, a PCP, and/or as part of the Prerequisite Control Program (e.g., using a supplier that can provide the ingredient without the hazard (see below and Table 3.4 in example of cookie dough) reduces the need to control it in the restaurant)). Remember all controls must be validated by the most current science and/or the FDA Food Code or be approved by the local health department if not validated.

   (a) Include the critical limit in each control including the CCP, the PCP, and/or the Prerequisite Control Program (e.g., a critical limit checked at the process of receiving is to check to ensure the raw cookie dough is from the approved supplier that used heated flour in the preparation of the ingredient where the hazard was controlled).

4. Define where each control will be monitored on a daily basis depending on where the hazards are most probable using the FSMS (Chap. 4) which will include verification by a CFPM.
5. Establish a periodic validation of the plan especially when any new menu item is added or deleted and when any ingredient change is made to a menu item/recipe that could remove the hazard or change where it is controlled best (e.g., at a supplier's facility).

Now let's practice this by demonstrating the design of a Process HACCP plan and a Prerequisite Control Program based on a fictitious restaurant concept called **Burger Circus** that only sells hamburgers, French fries (one type), drinks, and a raw cookie dough milk shake as its complete set of menu items. Using the steps described above, we first list all the menu items and ingredients used in each recipe that the restaurant concept serves to its customers (Table 3.4). In this same table, we then list probable biological, chemical, and physical hazards associated with each ingredient and also where each hazard is most probable to occur in the process (e.g., using defined source and hazards in Tables 3.1, 3.2, and 3.3). We can then complete a second table describing the hazard, where in the food preparation process it can be controlled and monitored at (at receiving, storing, preparing, cooking, cooling, reheating, hot holding, and serving), its control as a CCP/PCP, each critical limit, where to monitor each control in the food preparation processes, and what the corrective action should be when found not in compliance (Table 3.5). Recall that this

**Table 3.4** Burger Circus menu steps 1 thru 3 of Process HACCP plan design (hazard analysis) including hazards that will need to be controlled by the Prerequisite Program and PCPs

| Menu item | Ingredients used | Hazards | Where hazard most probable |
|---|---|---|---|
| **Hamburgers** | Raw ground beef | Shiga-toxin-producing *E. coli* | Supplier and during handling/cooking |
| | Wheat- and gluten-free buns | Wheat allergen | Supplier |
| | American cheese | *Listeria monocytogenes* or other dairy-related pathogens | Supplier |
| | Lettuce | Produce-associated pathogens like Shiga–toxin-producing *E. coli*, *Salmonella*, *Listeria*, and *Cyclospora* | Supplier and during preparation |
| | Tomatoes | *Salmonella* and Shiga-toxin-producing *E. coli* | Supplier and during preparation |
| | Pickles | Allergens | Supplier |
| | Condiments (ketchup, mayonnaise, mustard) | Allergens/physical hazards | Suppliers |
| **French fries** | Raw potatoes | Physical hazards such as bones, rocks, etc. and *Salmonella* | Supplier and during handling/cooking |
| **Drinks** | Prepared mix | Mold/allergens | Supplier |
| | Ice | Mold | During preparation of ice |
| **Milk shakes** | Milk shake base shelf stable | *Listeria monocytogenes* or other dairy-related pathogens | Supplier |
| | Cookie dough | Shiga-toxin-producing *E. coli*/physical hazards | Supplier |

**Table 3.5** Burger Circus menu steps 3 and 4 of Process HACCP plan design (hazard analysis) including hazards controlled by the Prerequisite Control Program and PCPs

| Food preparation process | Hazards/source | CCP/PCP/limit | Monitoring | Corrective action |
|---|---|---|---|---|
| **Receive** | Shiga-toxin-producing *E. coli* in raw ground beef | Approved supplier that controls Shiga–toxin-producing *E. coli* and performs microbiological testing for pathogens of final product according to USDA requirements | Check that product is from an approved supplier against current list | Reject product and source replacement |
| | Wheat allergen in gluten-free buns | Approved supplier | Check that product is from an approved supplier against current list | Reject product and source replacement |
| | *Listeria monocytogenes* or other dairy-related pathogens in American cheese | Approved supplier | Check that product is from an approved supplier against current list AND the product is received at 41 °F or below with no evidence of freeze/thaw at delivery | Reject product and source replacement |
| | Produce-associated pathogens like Shiga-toxin-producing *E. coli*, *Salmonella*, *Listeria*, and *Cyclospora* on lettuce | Approved supplier that has been audited and certified against produce safety standards such as GFSI | Check that product is from an approved supplier against current list AND the product is received at 41 °F or below | Reject product and source replacement |
| | *Salmonella* and Shiga-toxin-producing *E. coli* on tomatoes | Approved supplier that has been audited and certified against produce safety standards such as GFSI | Check that product is from an approved supplier against current list AND the product is received at 41 °F or below | Reject product and source replacement |

(continued)

**Table 3.5**  (continued)

| Food preparation process | Hazards/source | CCP/PCP/limit | Monitoring | Corrective action |
|---|---|---|---|---|
| | Allergens in pickles | Approved supplier | Check that product is from an approved supplier against current list | Reject product and source replacement |
| | Allergens/physical hazards in condiments (ketchup, mayonnaise, mustard) | Approved supplier | Check that product is from an approved supplier against current list | Reject product and source replacement |
| | Physical hazards such as bones, rocks, etc. and *Salmonella* in raw potatoes | Approved supplier that has been audited and certified against produce safety standards such as GFSI | Check that product is from an approved supplier against current list AND the product is received at 41 °F or below with no evidence of freeze/thaw at delivery | Reject product and source replacement |
| | Mold/allergens in prepared beverage mix | Approved supplier | Check that product is from an approved supplier against current list | Reject product and source replacement |
| | *Listeria monocytogenes* or other dairy-related pathogens in Milk shake base (not shelf stable) | Approved supplier that has been audited and certified against dairy safety standards such as GFSI | Check that product is from an approved supplier against current list AND the product is received at 41 °F or below with no evidence of freeze/thaw at delivery | Reject product and source replacement |
| | Shiga-toxin-producing *E. coli*/physical hazards in cookie dough | Approved supplier that has been audited and certified against produce safety standards such as GFSI | Check that product is from an approved supplier against current list AND the product is received at 41 °F or below with no evidence of freeze/thaw at delivery | Reject product and source replacement |

(continued)

**Table 3.5** (continued)

| Food preparation process | Hazards/source | CCP/PCP/limit | Monitoring | Corrective action |
|---|---|---|---|---|
| **Store** | Ground meat—Shiga-toxin-producing *E. coli* in raw ground beef proliferation; ; all produce except pickles + raw potatoes and cookie dough—Listeria monocytogenes, or other diary pathogens, Produce-associated pathogens, and shiga-toxin-producing *E. coli* proliferation | Keep product at 41 °F in walk-in refrigerator until prep | Check that raw ground beef and all other ingredients are held at 41 °F or below with no evidence of freeze/thaw at delivery | Separate bulk raw product or prepared patties or all other ingredients into smaller portions |
| **Prepare** | Prerequisite Control Programs—personal hygiene PCPs and cleaning and sanitation program; all produce ingredient hazards | Employee washing hands Proper foodservice glove use Raw ground beef patties prepared in separate area than RTE ingredients; all produce ingredients washed properly | Check to ensure employee washes hands before wearing gloves and uses single-use foodservice gloves when handling RTE ingredients including cooked patties. Ensure preparation of raw ground beef patties is separate from area where RTE cooked beef patties and other ingredients are used. Ensure produce is washed in a separate produce sink and spun dry | Have employee stop task, wash hands properly, and don new foodservice gloves. Discard any RTE ingredient/product prepared on food contact surface where raw beef was prepared and clean/sanitize the surface properly before further food prep; re-train employee if prior history requires; discard any unwashed produce if used to make final product and the final products. Wash produce if before |

(continued)

**Table 3.5** (continued)

| Food preparation process | Hazards/source | CCP/PCP/limit | Monitoring | Corrective action |
|---|---|---|---|---|
| **Cook** | Ground meat— Shiga-toxin- producing *E. coli* in raw ground beef | Cook ground beef to 155 °F | Check the internal beef patty temperature using a calibrated thermometer after a complete cook cycle | Continue cooking or recook the patty if removed from grill to 155 °F rapidly or discard the patty. Check other cooked patties to ensure equipment is cooking properly |
| | French fry potatoes— physical hazards such as bones, rocks, etc. and *Salmonella* in raw potatoes | Cook potatoes to 145 °F | Check the internal potato temperature using a calibrated thermometer after a complete cook cycle | |
| **Cool** | NA | NA | NA | NA |
| **Reheat** | NA | NA | NA | NA |
| **Hot hold** | Cooked hamburger patties | Hold product at 135 °F or higher | Check if the internal beef patty temperature is 135 °F or higher | Reheat the patty to 165 °F rapidly or discard the patty. Check other cooked patties |
| **Serve** | Prerequisite programs— personal hygiene PCPs | Employee washing hands Proper foodservice glove use | Check to ensure employee washes hands before wearing gloves and uses single-use foodservice gloves when handling RTE ingredients including cooked patties | Have employee stop task, wash hands properly, and don new foodservice gloves; re-train employee if prior history requires |

is not the FSMS by itself because this is the documented plan and Prerequisite Control Program Plan that must be developed into FSMS that can be executed in the day-to-day operations of the foodservice business (Chap. 4).

The hazard analysis step is made easiest by first checking for those already well known. The most common hazards in meats are well known such as the Shiga-toxin-producing *E. coli* in raw ground beef. This hazard should be controlled and monitored in four places because of potential failure of a control at a supplier or distributor level which are not under the direct management by the foodservice business, as follows:

1. An approved supplier whose ingredient meets USDA requirements is to test for Shiga-toxin-producing *E. coli* before the product is shipped.

2. The product is checked at receiving to ensure it is from approved source and received at the proper temperature (remember that Shiga-toxin-producing *E. coli* could still be present in raw ground beef and then grow to higher numbers when temperature abuse occurs).
3. During the preparation of raw hamburger patties, monitor to ensure separation, proper handling, and proper cleaning and sanitizing of food contact surfaces.
4. During the cooking process, monitor to ensure they are cooked at 155 °F, and during hot holding of cooked patties, monitor to ensure they are held at 135 °F.

You can also see that the restaurant sells hamburgers with gluten-free buns, and customers who may be allergic to wheat and/or have Crohn's disease aggravated by wheat/gluten would expect that this allergen/ingredient will not be in the bun. This hazard should be controlled and monitored in two places: at the approved supplier's facility and monitored at receiving (Table 3.5).

The hazard associated with the raw whole peeled potatoes seems more straight-forward as only a physical hazard seems likely and that is likely rare. Note because potatoes are harvested by machines, they often pick up other small foreign objects as the potatoes are pulled out of the soil (like a piece of a cow bone) that may be buried in the fields and can end up mixed in the final packaging of the product (that is a funny story I can share another time). Let's discuss a scenario where one ingredient like these potatoes used to make a current menu item (where only one hazard has been identified) might also be used at another time (e.g., a Limited Time Offer, LTO) where the hazards may change even though it is the same ingredient now used in a different recipe/food preparation process.

Suppose Burger Circus decided to sell baked potatoes using the same raw whole peeled potatoes ingredient used to cut and cook French Fries (Table 3.4). It may seem that the only hazard associated with using raw potatoes as an ingredient is a physical hazard described above. However, if the raw potatoes were used in another menu item, for example, baked potatoes or in a potato salad, then the hazard analy-sis would need to include additional hazards that should be controlled. Food safety issues associated with potatoes often involve prepared dishes, such as potato and other deli-style salads and baked potatoes. Potato salad was the likely source of a *Salmonella* gastroenteritis outbreak in 2009 (CDC 2010) and a Norovirus outbreak traced to an ill food handler (Chandler et al. 2000). *Bacillus cereus* poisoning has been associated with mashed potatoes (Palma 1985) and mashed potatoes made with raw milk were the suspected source of a *Staphylococcus aureus* outbreak in Norway (Jorgensen et al. 2005). Botulism has been linked to temperature abused potato salad (Brent et al. 1995) as well. Therefore you should always ensure a com-plete hazard analysis of each food ingredient based on how the ingredient will be used to prepare the final product including any future use of that ingredient in dif-ferent recipes if feasible. It may not be feasible to know which new menu items will be introduced in the future at the time of the development of the Process HACCP plan, so the plan and Prerequisite Control Program should be reevaluated anytime the menu is changed to ensure all hazards are defined.

Remember that all specialized menu items when conducting certain specialized food preparation processes like reduced oxygen packaging of TCS foods may need to be submitted for approval to the local health department as part of the Process

HACCP plan development. The hazard analysis will need to define the specialized control, and be monitored during each preparation of that menu item to include documentation that the controls are in place for all such products prepared in a restaurant. The local health department will likely inspect a foodservice business and request to see documented monitoring data for all products prepared under an allowance of variance to prepare specialized menu items. This can easily be accomplished using a FSMS defined in Chap. 4 already in place. However at the time of this book authorship, Process HACCP and FSMS are not required nor regulated in foodservice businesses in the United States (this may soon change) but HACCP plans for certain specialized food preparation processes are required and regulated.

Once you have completed the Process HACCP plan using examples like in Tables 3.4 and 3.5 using all of the menu items you sell to customers, you are ready to design the FSMS you need to ensure execution of the Process HACCP plan and Prerequisite Control Program required to prevent foodborne illnesses and outbreaks in the restaurant.

# References

Brent J et al (1995) Botulism from potato salad. Dairy Food Environ Sanit 15(7):420–422

CDC (2010) Multiple- serotype *Salmonella* gastroenteritis outbreak after a reception, Connecticut, 2009. MMWR 59(34):1093–1097

CDC (2019) Say no to raw dough. Content source: Centers for Disease Control and Prevention, National Center for Emerging and Zoonotic Infectious Diseases (NCEZID), Division of Foodborne, Waterborne, and Environmental Diseases (DFWED).https://www.cdc.gov/food-safety/communication/no-raw-dough.html

Chandler B et al (2000) Outbreaks of Norwalk-like viral gastroenteritis – Alaska and Wisconsin, 1999. Morb Mortal Wkly Rep 49(10):207–211

Food and Drug Administration (FDA) (2006) Managing food safety: a manual for the voluntary use of HACCP principles for operators of food service and retail establishments. U.S. Department of Health and Human Services, Public Health Service. https://www.fda.gov/food/hazard-anal-ysis-critical-control-point-haccp/managing-food-safety-manual-voluntary-use-haccp-princi-ples-operators-food-service-and-retail

Food and Drug Administration (FDA) (2017) FDA food code. U. S. Department of Health and Human Services, Public Health Service, Washington, DC. https://www.fda.gov/Food/GuidanceRegulation/RetailFoodProtection/FoodCode/ucm595139.htm

Jorgensen H et al (2005) An outbreak of staphylococcal food poisoning caused by enterotoxin H in mashed potato made with raw milk. FEMS Microbiol Lett 252:267–272

King H (2013) Food safety management: implementing a food safety program in a food retail business. Springer, New York

King H, Ades G (2015) Hazard Analysis and Risk-Based Preventive Controls (HARPC): the new GMP for food manufacturing. Food Safety Magazine, Oct/Nov. http://www.foodsafetymaga-zine.com/magazine-archive1/octobernovember-2015/hazard-analysis-and-risk-based-preven-tive-controls-harpc-the-new-gmp-for-food-manufacturing/

King H, Bedale W (2017) Hazard analysis and risk-based preventive controls: improving food safety in human food manufacturing for food businesses. Elsevier, London

King H, Bedale W (2018) Managing food allergens in retail quick-service restaurants. In: Food allergens; best practices for assessing, managing, and communicating risks. Springer, Cham

King H, Michaels B (2019) The need for a glove-use management system in retail foodservice. June/July 2019. See: https://www.foodsafetymagazine.com/magazine-archive1/junejuly-2019/the-need-for-a-glove-use-management-system-in-retail-foodservice/

Mortimore S, Wallace C (2013) HACCP: a practical approach. Springer, New York

Palma A (1985) Mashed potatoes and Bacillus cereus poisoning. Ristorazione Collettiva 10(3):117–118

SurakJG (2009) The evolution of HACCP. Food Quality and Safety. Feb/Mar issue. Wiley Company

Warren CM et al (2019) Prevalence and severity of sesame allergy in the United States. JAMA Netw Open 2(8):e199144. https://doi.org/10.1001/jamanetworkopen.2019.9144

# Chapter 4
# Design of Food Safety Management Systems Using the Process HACCP Plan and Prerequisite Control Program

The best Food Safety Management System(s) (FSMS) to enable AMC start with a foundation of the Process HACCP plan with a well-defined Prerequisite Control Program specific to the foodservice businesses' menu. The function of FSMS is to ensure daily execution of all controls necessary to eliminate all biological, chemical, and physical hazards that lead to foodborne illnesses and outbreaks (see Chap. 2). The Food Safety Management System should be developed based on the Process HACCP plan and the Prerequisite Control Program, and establish the procedures (how to perform assessment), monitoring method (what tools will be used), and documentation of the results that show control of all hazards.

There must be a clear plan of training to enable food handler employees to be trained to perform the food preparation processes (from receiving to service) and all standard operating procedures (SOPs) safely when not being monitored. There needs to be a training program at a Certified Food Protection Manager (CFPM)-based level that also enables a manager (the person in charge at each shift of operations) to execute the FSMS to monitor safe food preparation and make appropriate and effective corrective actions when needed (see Chap. 5). The importance of selecting which food safety training to use especially for food handlers will also be evident when it is used as a corrective action, for example, *re-training* an employee on personal hygiene controls. When observed repeatedly not in compliance, it can be an effective corrective action. Once either the food handler/employee does not execute any part of the safe food preparation processes and controls or the managers stop using the FSMS to manage them, the hazards will be more probable leading to a foodborne disease illness.

A digital technology is necessary to ensure the FSMS is "active" and not passive on paper to ensure proper management of SOPs, visibility of results to the foodservice business, and immediate execution of corrective actions to control hazards (see Chap. 7). The Food Safety Management System(s) will be more effective when the facilities, especially the kitchen and foodservice areas, are designed for the safe storage and preparation of foods (e.g., areas of separation for raw protein and/or allergen-containing food prep; see Chap. 6). Verification that the FSMS are being

H. King, *Food Safety Management Systems*, Food Microbiology and Food Safety, https://doi.org/10.1007/978-3-030-44735-9_4

implemented correctly is an important function of the owner(s)/general manager of a restaurant (i.e., the person all the managers report to) including follow-up of all root-cause assessments when controls fail. Finally, the validation function is important to ensure all of the FSMS are correct and current, any updates in design are made, change in equipment use or facility design is updated, and training of employees (including any re-training) is current that will control all of the hazards identified from the most current menu being prepared and served to customers.

## Food Safety Management Systems Vs. System

Throughout this book, I have referenced the need for Food Safety Management System(s) (using FSMS) rather than just a system based on the FDA definition of a FSMS. This is because there are numerous food safety hazards that can contaminate food throughout the supply chain to the customer by numerous contributing factors, and they cannot all be controlled by just one Food Safety Management System. However the primary FSMS should start with the Process HACCP plan based on the menu and include the hazards defined in the Prerequisite Control Program (Chap.3), and this must be documented in order to enable training of employees on it and execution of it in each foodservice location. If all of the identified hazards can be controlled by one FSMS because, for example, many of the controls are managed outside of the business and the business does not cook food, then one FSMA may suffice. This will be evident after you have developed the Process HACCP plan and Prerequisite Control Program.

## The Case for Food Safety Management Systems

A Food Safety Management System is defined by the FDA as a specific set of actions that include procedures, training, and monitoring (also designated P, T, and M; see below) to help a foodservice establishment achieve Active Managerial Control of foodborne disease risk factors (FDA 2018). The fact that many restaurants do not use and/or have inadequate FSMS is believed to contribute to a large number of foodborne disease illnesses and outbreaks (Luning et al. 2008), and the missing foundation or poor implementation of HACCP to prevent hazards has been described as the primary contributing factor (Luning et al. 2009). When a restaurant only uses checklist to monitor food safety risk most often with a focus only on temperatures, it is not using an FSMS, in my opinion, and likely will miss identifying hazards and their important controls necessary to prevent foodborne disease illnesses.

The two most important components necessary to design and implement an effective FSMS to prevent foodborne illness are:

**Table 4.1** Foodborne illness risk factors and the associated primary data items examined in the study

| Foodborne illness risk factor | Associated primary data item numbers and description |
|---|---|
| Poor personal hygiene | Data item #1—Employees practice proper hand washing |
| | Data item #2—Employees do not contact ready-to-eat foods with bare hands |
| Contaminated equipment/ protection from contamination | Data item #3—Food is protected from cross-contamination during storage, preparation, and display |
| | Data item #4—Food contact surfaces are properly cleaned and sanitized |
| Improper holding time/ temperature | Data item #5—Foods requiring refrigeration are held at the proper temperature |
| | Data item #6—Foods displayed or stored hot are held at the proper temperature |
| | Data item #7—Foods are cooled properly |
| | Data item #8—Refrigerated, ready-to-eat foods are properly date marked and discarded within 7 days of preparation or opening |
| Inadequate cooking | Data item #9—Raw animal foods are cooked to required temperatures |
| | Data item #10—Cooked foods are reheated to required temperatures |

Source: FDA (2018)

1. The Process HACCP plan for the foodservice business
2. The Prerequisite Control Program for the foodservice business

Together, both are necessary to the effective execution of the foodservice businesses' FSMS to ensure control of all hazards associated with foodborne illnesses and outbreaks. The execution of the FSMS by a CFPM ensures the knowledge necessary to perform corrective actions when controls are monitored and not in compliance but also root-cause assessment of how the control was lost (if not immediately evident) is important; such as, why a food is not cooking properly (equipment failure, employee failure, etc.) to enable a corrective action. In fact, when a CFPM performs daily assessments of controls using FSMS, they and the empolyees are continually exposed to the Process HACCP plan and Prerequisite Control Program requirements that will improve their food safety management skills as well.

Direct evidence of the importance of a CFPM and FSMS in the prevention of foodborne illness risk factors in restaurants was shown by the FDA risk assessment on the occurrence of foodborne illness risk factors in fast-food and full-service restaurants (FDA 2018). The study actually started in 2013, and the actual restaurant audits of over 396 full-service restaurants and 425 fast-food restaurants were completed in 2014. As with some government-managed research programs, an unfortunate delay occurred before these data were published. Nonetheless, the data is important to the industry as it directly shows (via FDA auditors that performed an audit in each of these 821 restaurants across the United States) the impact of a CFPM and FSMS on four of the five foodborne disease/illness risk factors

(Table 4.1; note it was likely difficult to audit safe source of food controls—the fifth risk factor—at the time of the audits).

To assess the presence and impact of a CFPM and FSMS in each of the 821 restaurants, the FDA audited the presence and effect of a CFPM in each restaurant. They also evaluated the presence and effect of the three primary components (see below) of an FSMS measuring the evidence of a partial vs. fully documented FSMS and how each of the 821 restaurants were currently controlling the risk factors at the time of the audit (in Table 4.1), rating each restaurant's ability to control each of the four risk factors (see Chap. 3 to review all five risk factors) against the presence/absence of a CFPM and FSMS:

- **Procedures (P):** A defined set of actions adopted by foodservice management for accomplishing a task in a way that minimizes food safety risks
- **Training (T):** The process of management's informing employees of the food safety procedures within the restaurant and teaching employees how to carry them out
- **Monitoring (M):** Routine observations and measurements conducted to determine if food safety procedures are being followed and maintained

The FDA rated each restaurant's level of FSMS design and execution according to a PTM score (presence or absence of P, T, and M) and established the definitions of compliance as; nonexistent FSMS (no system in place), underdeveloped FSMS (system in development but many gaps), well-developed FSMS (system is complete, consistent, and partly documented), and well-developed and documented FSMS (system is complete, consistent, and written). A well-developed and documented FSMS would be expected (as part of the hypothesis as to what the FDA predicted would be found) to have the least amount of foodborne illness risk factors present in the foodservice operations.

The FDA FSMS risk factor study, as expected (see Chap. 2 with over 60% of all foodborne disease outbreaks in the United States due to restaurants year to year), showed that only 10.26% of the 425 fast-food restaurants and only 2.78% of the full-service restaurants had a well-developed and documented Food Safety Management System (i.e., a fully operational and executed FSMS that would control the hazards that cause foodborne illnesses and outbreaks). The good news in the report was 76.37% of the 425 fast-food restaurants and 65.91% of the full-service restaurants had at least an underdeveloped/well-developed but not documented Food Safety Management System. Interestingly, the top three contributing factors (see Appendix A) in all the restaurants audited were the same in both fast-food and full-service restaurants (Table 4.2): (1) foods requiring refrigeration not at the proper temperature, (2) employees not practicing proper hand washing, and (3) foods not cooled properly. Several other contributing factors were observed at high percentages as well most likely because few of the 821 restaurants were found to have a well-developed and documented FSMS.

**Table 4.2** Primary data items out-of-compliance during the FDA inspections of restaurants, descending order of percentage

| Data item | Fast-food restaurants Data item description | % OUT | Data item | Full-service restaurants Data item description | % OUT |
|---|---|---|---|---|---|
| 5 | Foods requiring refrigeration are held at proper temperature | 68.24 | 5 | Foods requiring refrigeration are held at proper temperature | 86.11 |
| 1 | Employees practice proper hand washing | 65.64 | 1 | Employees practice proper hand washing | 82.40 |
| 7 | Foods are cooled properly | 49.42 | 7 | Foods are cooled properly | 71.79 |
| 4 | Food contact surfaces are properly cleaned and sanitized | 40.94 | 8 | Refrigerated, ready-to-eat foods are properly date marked and discarded within 7 days of preparation or opening | 70.65 |
| 3 | Food is protected from cross-contamination during storage, preparation, and display | 36.94 | 3 | Food is protected from cross-contamination during storage, preparation, and display | 66.92 |
| | Refrigerated, ready-to-eat foods are properly date marked and discarded within 7 days of preparation or opening | 32.09 | 4 | Food contact surfaces are properly cleaned and sanitized | 62.12 |
| 6 | Foods displayed or stored hot are held at proper temperature | 23.95 | 10 | Cooked foods are reheated to required temperatures | 36.64 |
| 10 | Cooked foods are reheated to required temperatures | 16.03 | 6 | Foods displayed or stored hot are held at proper temperature | 34.73 |
| 2 | Employees do not contact ready-to-eat foods with bare hands | 2.47 | 2 | Employees do not contact ready-to-eat foods with bare hands | 33.59 |
| 9 | Raw animal foods are cooked to required temperatures | 10.65 | 9 | Raw animal foods are cooked to required temperatures | 21.05 |

Source: FDA (2018)

Restaurants associated with a multi-location (chain) restaurant business had fewer contributing factors in most of the risk factor categories during the audits and were most likely to have a more developed FSMS compared to single-location independent (non-chain-associated) restaurant businesses. Overall, both fast-food and full-service restaurants were audited and found to have the most control over having employees not contact RTE foods with bare hands (with gloves) and cooking raw animal foods to the proper temperature. These observations correlated to the findings that a large percentage of both restaurant types had at least an underdeveloped/well-developed but not documented Food Safety Management System and, in my personal opinion, were likely related to the presence of a CFPM as the person in charge being present at the time of the FDA audit (64.47% percent of fast-food restaurants and 58.08% of the full-service restaurants had a CFPM as the person in charge). However, having a CFPM and even executing some controls when a CFPM is present as the person in charge is not enough to prevent a foodborne illnesses or outbreaks in a restaurant.

Not having an FSMS was the strongest predictor of all items being out-of-compliance in both fast-food and full-service restaurants. The restaurants with well-developed FSMS had significantly fewer food safety behaviors/practices than did those with less developed FSMS. For example, fast-food restaurants without an FSMS (nonexistent) averaged 4.5 of the contributing factor controls out-of-compliance while fast-food restaurants with a well-developed FSMS averaged fewer than 1.7 controls out-of-compliance. The FDA analysis of the data from all restaurants audited showed that an FSMS and not the presence of a CFPM alone predicted compliance with these food safety behaviors/practices. The FDA study did not look at the controls of all contributing factors that lead to risk factors as a definition of a FSMS (e.g., working sick employees, proper cleaning and sanitation, foods from unsafe sources, allergen management, and other Prerequisite Control Program-based FSMS elements; see below) nor did they audit for the type of plan a FSMS was based on such as a critical Process HACCP-based FSMS prescribed by the FDA. Thus, it might be expected that with a Process HACCP-based FSMS in place (as discussed in this chapter), the restaurants audited by the FDA would have performed significantly better in controlling the contributing factors and preventing foodborne illness risk factors.

## Prerequisite Control Program for the Control of Hazards in Food Safety Management Systems

As discussed in Chaps. 1 and 3, when a Prerequisite Control Program (i.e., the policy) is defined and executed with the Process HACCP plan, the FSMS are more effective and more comprehensive in controlling all of the hazards associated with the menu a foodservice business prepares and sells. While the Process HACCP plan is developed to identify hazards (via the hazard analysis) and their controls, the Prerequisite Control Program is primarily designed to identify known controls of contributing factors necessary to ensure the safe preparation of foods, for example, when employee training is defined (to ensure each employee knows what they are expected to do and how they will be held accountable to it via the FSMS), allergen control, chemical safety, pest control, cross-contamination risk critical to the safe food preparation processes, and ensuring a safe source of ingredients and products from the restaurant's suppliers.

In a multi-location (chain) foodservice business, a Prerequisite Control Program may be managed by business functions within different departments of the organization such as a supplier food safety management function in the Supply Chain/Procurement Department, pest control and the training function in the Operations Department, equipment and facility management in the Facilities Department, etc. Nevertheless, a Prerequisite Control Program is essential to the FSMS used day to day in the foodservice location even if many of the functions are contracted to vendors to perform by a single-location independent business. The minimum recommended components of a Prerequisite Control Program that will be needed in the FSMS include:

- Methods to ensure equipment maintenance

  - Equipment to ensure temperature measuring devices are calibrated.
  - Cooking equipment is calibrated, and hot and cold holding equipment provides the correct temperature.
  - Refrigeration and freezer equipment provide the correct environmental temperature.
  - Ware washing equipment are operating according to manufacturer's specifications.

- Methods to ensure allergen management in food preparation and storage
- Methods to ensure safe chemical use and storage around foods and employee safety that meet Occupational Safety and Health Administration (OSHA) requirements for foodservice businesses (see https://www.osha.gov/SLTC/restaurant/)
- Methods to ensure safe water use for food and in the manufacture of ice
- An effective pest prevention program to prevent pest infestations
- Methods to ensure no bare-hand contact with any ready-to-eat (RTE) food to prevent the cross-contamination of foods from hands
- Methods to ensure proper hand washing to prevent the cross-contamination of foods from hands, including when wearing gloves
- Methods to ensure restriction and exclusion of sick employees who have known signs, symptoms, or diagnosis of foodborne illnesses to prevent the cross-contamination of foods from hands including cuts and burns (as part of the health policy; see Chap. 4)
- Personal hygiene requirements of employees working around food preparation (clean clothing, hair restraints, eating/smoking/drinking restrictions, jewelry restrictions)
- A cleaning and sanitation program for the direct prevention of cross-contamination of ready-to-eat foods by raw animal foods, clean and sanitized food contact surfaces, cutting boards, dish washing equipment, utensils, aprons, etc.
- Methods to ensure safe source of foods—using only food safety ingredient and food suppliers (e.g., GFSI- and FSMA-compliant supplier food manufacturing facilities) to ensure a safe source of food, and food packaging is safe to serve food (e.g., meets FDA food packaging in contact with food requirements)
- Methods to ensure ingredients are not used past their safe expiration date using FIFO
- Methods to ensure an ingredient/food product in use has not been recalled by the FDA nor by a CDC "do not consume" communications to ensure the restaurant is alerted to when the FDA communicates not to serve an ingredient/food product or the CDC communicates not to consume an ingredient/food product

# A Food Safety Management System Based on the Process HACCP Plan and Prerequisite Control Program

Now, in order to establish the FSMS once the specific Process HACCP and Prerequisite Control Program have been designed (see Chap. 3), you need to determine how each control will be monitored and where and when to monitor them to ensure all of the foodborne illness risk factors are managed at each shift of operations. First and most importantly, the monitoring of all controls should be designed according to when they are most probable (which may be most likely during active food preparation processes) in the flow of food (see Fig. 4.1, from the steps receiving to serving). This is why we focus on the Process HACCP plan as the foundation of the FSMS; where if the control is found non-compliant, the manager may make corrective action immediately to ensure the affected food is safe before service.

For example, one of the important controls that is necessary to ensure food is from a safe source (a risk factor that has numerous hazards and contributing factors associated with it) that meets the specifications the food supplier should have followed (as part of the Prerequisite Control Program) is to monitor all foods received into the restaurant at the time of delivery. This is important so that if a product/ingredient is not from an approved vendor and/or compliant to temperature and packaging requirements, it can be rejected and not received and stored in the restaurant. Of course, you should also monitor for this risk factor during the storing process as well (even when there is no food preparation occurring) to check if all ingredients and foods are from approved sources and have not passed their date of expiration for safety. In this example, you would include monitoring storage of foods for the proper control using date of expiration marking/labeling to ensure all ingredients and in-house prepared foods (those that must never be sold past 7 days post-preparation) are present.

Using the example format in Table 3.5 from Chap. 3 where the Process HACCP plan defined the CCP/PPC/limit, monitoring, and corrective actions for the fictitious **Burger Circus** restaurant and Table 3.4 from Chap. 3 where the example format defined where the hazard is most probable, we can now design the FSMS to properly monitor all controls during operations (Fig. 4.1). Since this example of FSMS is based on the Process HACCP plan developed in Chap. 3 for the fictitious restaurant called Burger Circus (Fig. 3.5, Chap.3), the CFPM and his/her designated employees would perform assessments of these items at a time when the hazards are most likely to be prevented by the control performed during each active food preparation process of:

1. **Receiving** (especially important when products are sourced from suppliers that have been specified to control hazard(s) (such as Shiga-toxin-producing *E. coli* in ground beef)); the control here is to ensure only these approved suppliers' ingredients/products are delivered to the restaurant. If raw eggs were used in the menu (they are not in this fictitious Burger Circus restaurant), the interior of the delivery vehicle must be checked to ensure it is at or below 45 °F at all times during transportation.

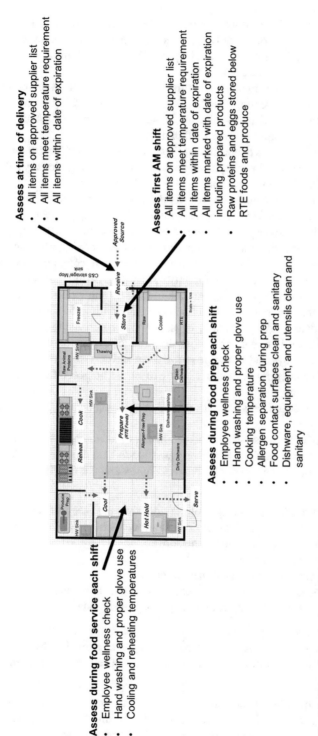

**Assess at time of delivery**
- All items on approved supplier list
- All items meet temperature requirement
- All items within date of expiration

**Assess first AM shift**
- All items on approved supplier list
- All items meet temperature requirement
- All items within date of expiration
- All items marked with date of expiration including prepared products
- Raw proteins and eggs stored below RTE foods and produce

**Assess during food prep each shift**
- Employee wellness check
- Hand washing and proper glove use
- Cooking temperature
- Allergen separation during prep
- Food contact surfaces clean and sanitary
- Dishware, equipment, and utensils clean and sanitary

**Assess during food service each shift**
- Employee wellness check
- Hand washing and proper glove use
- Cooling and reheating temperatures

**Fig. 4.1** Example monitoring of the controls identified in the Process HACCP plan using an FSMS along the flow of food preparation processes designated as CCPs and PCPs. See Chap. 6 for more details on how facility can be defined to enable FSMS

(a) Inspect the delivery truck when it arrives to ensure that it is clean, free of putrid odors, and organized to prevent cross-contamination. Be sure refrigerated foods are delivered on a refrigerated truck. Check the interior temperature of refrigerated trucks.

(b) Confirm vendor name, day and time of delivery, as well as driver's identification before accepting delivery. If driver's name is different from what is indicated on the delivery schedule, contact the vendor/distributor immediately, and do not receive the food item until resolved.

(c) Check frozen foods to ensure that they are all frozen solid and show no signs of thawing and refreezing, such as the presence of large ice crystals or liquids on the bottom of cartons.

(d) For packaged products, insert a food thermometer between two packages being careful not to puncture the wrapper. If the temperature exceeds 41 °F, it may be necessary to take the internal temperature before accepting the product unless there is a record of time out-of-compliance via a digital tracking sensor. Note: in order to reduce the need for rejection of food products that may be above 41 °F but for only a short time period (under 4 h), you can use temperature sensors called data loggers to show time and temperature during perishable food distribution especially during the summer season.

(e) Check date marking of milk, eggs, and other perishable goods to ensure enough time is remaining before the product expiration date on the label (and/or best by date) to enable preparation and service.

(f) Check the integrity of food packaging to ensure no pest infestation or rodent chew-through.

(g) Check the cleanliness of crates and other shipping containers before accepting products.

(h) **Corrective actions**—Reject all products or place in area with "do not use" signage for disposition if delivery was made as a drop-key service (where the product is delivered after business hours and not checked until the next day):

(i) Frozen foods with signs of previous thawing (keep frozen)

(ii) Cans that have signs of deterioration, such as swollen sides or ends, flawed seals or seams, dents, or rust

(iii) Punctured packages

(iv) Packaging with out-of-date products

(v) Foods that are out of safe temperature zone or deemed unacceptable by the established rejection policy (keep refrigerated)

2. **Storing** (temperature is critical and can be effective here to reduce growth (proliferation) of bacteria like *Listeria monocytogenes* in cheeses and RTE produce like lettuce and tomatoes, as it can grow in refrigerated ingredients and products):

(a) Check the internal temperature of all refrigerated and frozen products to ensure below 41 °F or 32 °F, respectively. Ensure the probe is cleaned (due to allergen cross-contact risk) and sanitized between use.

(b) Check to ensure no product in storage has passed its date of expiration, and label those close to a date of expiration to be used next according to FIFO.

(c) **Corrective actions**—If not longer than 4 hours, separate bulk raw products or RTE prepared foods into smaller portions, and recheck product temperatures. Alternatively, move product to freezer for a short period until product reaches proper temperature.

3. **Preparing** (because working sick employees can contaminate foods during preparation, it is important to perform a wellness check here as well (as should be done at the beginning of every shift perhaps at the time when an employee will clock-in for their shift). Note: see the most current FDA Food Code for additional history of exposure and past illness wellness checks, and use the FDA's *Retail Food Protection: Employee Health and Personal Hygiene Handbook* (see https://www.fda.gov/food/retail-food-industryregulatory-assistance-training/retail-food-protection-employee-health-and-personal-hygiene-handbook) as reference to develop your employee health policy. You should also monitor both for any updates on required foodborne illnesses (those transmitted through food) including FDA recommended exclusion and restriction from work requirements. A free app available at emergiprotect.com can also be used to help learn how to screen employees for illnesses (called a wellness check).

(a) Ask each employee preparing food at the time of the assessment if they have any of these symptoms:

  (i)   Diarrhea
  (ii)  Vomiting in the past 24 h
  (iii) Fever with sore throat
  (iv)  Jaundice (yellowing of the skin or eyes)
  (v)   Open wounds on hands/arms from cuts or burns

(b) If any employee confirms they have or have had any of these symptoms, then ask employee if they have been to the doctor and have been diagnosed with any of these illnesses:

  (i)   Norovirus
  (ii)  *Hepatitis* A
  (iii) *Shigella* spp.
  (iv)  Shiga-toxin-producing *E. coli*
  (v)   Typhoid fever (caused by *Salmonella* Typhi)
  (vi)  *Salmonella* (non-typhoidal)

(c) Check to ensure no employee that has been excluded from work is actually working during the shift (see corrective action below) by checking the "sick log" or other means to ensure sick employee does not return to work before allowed.

(d) Check to ensure employees are washing hands properly and at the appropriate times especially before and after glove use and are wearing gloves properly while preparing any RTE foods. Improper glove use by food handlers is a contributing factor to cross-contamination of foods (see King and Michaels (2019)).

(e) Check to ensure all food contact surfaces are clean and sanitary and all raw proteins are being prepared separately from any RTE foods. Ensure the proper detergent/cleaner is being used, and the proper sanitizer that will kill all foodborne pathogens including viruses like Norovirus and Hepatitis A is used. Also ensure reusable towels are stored and used properly (see below). These SOPs are part of the Prerequisite Control Program and should not be assumed but managed by FSMS.

(f) Check to ensure all food prep equipment and utensils are clean and sanitary before being used to prepare and hold foods. Ensure dirty dishware is not stored near/above clean dishware, and cleaned dishware is not stacked wet.

(g) Check to ensure ingredients or foods with allergens are being prepared separately from any foods that do not have allergens.

(h) **Corrective actions**:

(i) Exclude or restrict any employee from work that confirms any of the symptoms or illnesses for the proper exclusion period (see FDA's *Retail Food Protection: Employee Health and Personal Hygiene Handbook* for the correct procedures and times for exclusion/restriction, https://www.fda.gov/food/retail-food-industryregulatory-assistance-training/retail-food-protection-employee-health-and-personal-hygiene-handbook). If an employee has been excluded from work due to (a) or (b) above, they should be placed on a "sick log" or other means to prohibit them from clocking into a shift that can also be monitored by different management during any shift. Follow the recommendations in the handbook described above for how long an employee should be restricted from work with different symptoms vs. diagnosed illnesses as some are within 24–72 h while others may be many days (e.g., Hepatitis A).

(ii) Discard any RTE food that has been touched with bare hands or gloved hands where gloves have torn surface. Have employee remove gloves, wash hands properly, and don new foodservice gloves before handling foods.

(iii) Discard any RTE food that has come into contact with a food contact surface used to prepare raw animal protein at the same time (i.e., has not been properly cleaned and sanitized between use). Have employee reclean and sanitize any food contact surface and/or equipment or utensils.

(iv) Discard any non-allergen-containing food(s) that have come into contact with a food contact surface used to prepare ingredients with allergens at the same time (i.e., has not been properly cleaned and sanitized between use).

4. **Cooking**:

(a) Check the cooking process to ensure the products (in this case for Burger Circus, ground beef at 155 °F but if poultry it would be 165 °F and if pork or fish 145 °F) are cooked properly.

(b) Check to ensure the frozen raw French Fries prepared from raw whole potatoes do not have any physical hazards. If used for baked potatoes or as ingredient for other prepared food products (as discussed on Chap. 3), check to ensure the product is cooked to 145 °F.

(c) **Corrective actions**—Because the quality of many food products may be affected by recooking time and temperature, it is recommended that each be recooked for the minimum time required to reach the temperature described in (a) above. If the product does not reach the proper internal temperature, it must be discarded and the root cause of the cooking process determined.

5. **Cooling:**

(a) Check time on product being cooled down, and ensure temperature is below 70 °F after 4 h or 41 °F if after 6 h total.

(b) **Corrective actions**—If product is above 70 °F and it has not been longer than 4 h or if product is below 70 °F but higher than 41 °F and it has not been more than 2 h, you can reduce the volume or placement of product in storage containers to increase the surface area of the food or place in freezer (monitor to avoid freezing and move to refrigeration) and recheck temperature, and note time it is at or below 41 °F. If product is found to be at a temperature above 41 °F for over 6 h, discard the food, and reevaluate the methods of cooling products.

6. **Reheating:**

(a) Check product being reheated to ensure temperature is at or above 165 °F.

(b) **Corrective actions**—Continue to reheat the product until the temperature reaches 165 °F. If product quality is impacted or product does not reach this temperature, you should discard the product and determine root cause before reheating other products.

7. **Hot or cold holding:**

(a) Check product being held for service or prep, and ensure temperature is at or below 41 °F if cold holding or 135 °F if hot holding.

(b) **Corrective actions**—If product is above 41 °F but not 135 °F or below 135 °F but above 41 °F (i.e., the temperature danger zone: 41 °F to 135 °F), then re-cool or reheat the product according to number 5 or 6 (above), respectively. If product has been in the temperature danger zone for more than 4 h, discard the food, and determine root cause before reheating other products.

8. **Serving:**

(a) Check to ensure employees are washing hands before glove use and are wearing gloves properly while preparing any RTE foods.

(b) Check to ensure ice machine and service is clean and sanitary; ice is being scooped using clean and sanitary ice scoops (not cups).

(c) **Corrective actions**—Discard any RTE food that has been touched with bare hands or gloved hands where gloves have torn surface. Have employee remove gloves, wash hands properly, and don new foodservice gloves before handling foods. Clean and sanitize the ice machine equipment and use clean and sanitized ice scoops.

This FSMS performed daily and/or at each shift in parts (e.g., while food process is happening) can ensure that each of the controls of the defined hazards is in place at the time when the hazard is best controlled in the restaurant operations, and this will significantly reduce the risk of a foodborne illness from the restaurant (not just at this fictitious restaurant called Burger Circus). Remember that many of the hazards related to supplier defects (e.g., *E. coli* O157 in ground beef) are oftentimes required by the USDA and FDA to be controlled at the supplier's facility/business which should be documented in the Prerequisite Control Program for supplier food safety management via a FSMS (e.g., where the product is required to be tested to ensure it is free of detectable Shiga-toxin-producing *E. coli*). However, a food manufacturer can and does make mistakes during processing or there could be temperature abuse during transportation and distribution of the product after it has left the manufacturer that would enable growth (proliferation) of *E. coli* to higher numbers in the ground beef; thus, multiple controls are necessary to fully prevent this one hazard in food. By placing additional controls at the time of receiving (approved source and temperature compliance), storing (temperature compliance and storage of raw ground beef below RTE foods), cooking the hamburgers to 155 °F, holding the cooked patties at 135 °F or higher, and preparing food (e.g., not preparing RTE hamburgers near raw ground beef preparation) which are each only under the control of the restaurant business, the risk of this one hazard can be completely controlled.

The most important value of the FSMS is when the restaurant CFPM performs these assessments of the controls at each shift of the operations to ensure maximum prevention of foodborne illness because of the many variables below that will change operations in a restaurant from shift to shift (for example only, not limited to these):

- New employee begins a shift without time to train them on the basics of food safety so expected knowledge of hand washing, glove use, not working sick, proper food prep, etc. is not present.
- Trained employee calls in sick and restaurant does not have the necessary staff to ensure food safety during operations.
- High turnover rate of employees vs. high volume of product sales.
- Manager calls in sick out and thus current person in charge of restaurant does not have the knowledge to manage food safety controls.
- New menu items being sold or change in the menu.
- Equipment fails (cooker or dish washer) and not available for use or repair persons in the restaurant working on the equipment and around RTE foods.

- Large volume of customers or catering food prep not expected-increasing food prep needs pushing food prep into areas of kitchen reserved for raw animal protein prep.
- Large volume of third-party delivery pick-ups or catering orders that must be prepared on limited space in the kitchen.
- Key-drop deliveries where products are delivered at night without supervision by the business.
- Delivery truck arrives late or during busy day part or during hot summer day staging pallets of food outside.
- Notice of ingredient or food product withdrawal or recall changing available ingredients to make recipes.
- New complex recipe items are added (e.g., LTOs) to the menu.

Because there are also numerous contributing factors to the foodborne illness risk that are encountered in a restaurant (see Chaps. 1, 2, and Appendix A), the foodservice business must also establish additional FSMS based on the Prerequisite Control Program that may not routinely need to be checked during each of the food preparation processes (many are, of course, like hand washing and glove use during food preparation), but each has hazards and contributing factors that, if not monitored by a CFPM, may introduce these hazards into the foodservice facility. It isn't always necessary to monitor the controls for these on a daily basis (for many of Prerequisite Control Program components) as long as they are firmly established and function via a FSMS. Likewise, some may require monitoring outside of the restaurant, for example, a corporate FSMS for supply chain food safety management which is likely monitored at the corporate level by the food safety department by establishing supplier specifications for all ingredients and products sourced for its restaurants, certifications, and third-party audits (see below). The important thing is that someone is managing and monitoring them on a regular basis to ensure the Prerequisite Control Program is active.

## Other Food Safety Management Systems Based on the Prerequisite Control Program

The Process HACCP plan and the resultant FSMS described above are developed with the assumption that during the food preparation processes, all of the Prerequisite Control Program is in place; for example, all surfaces and utensils are cleaned and sanitized properly before food preparation processes are performed including for the safe storage of food. If one component of the Prerequisite Control Program is not developed and implemented into an FSMS such as a cleaning and sanitation FSMS, it could be the one contributing factor that leads to the largest foodborne disease illness outbreak even though all of the other hazards determined by the Process HACCP plan are controlled. Suppose employees are assessed and found to be preparing the food properly (washing hands, wearing gloves, preparing RTE food ingredients like chicken salad on clean and sanitized surfaces, etc.) according to the Process HACCP plan, as this FSMS is used to check to ensure all the CCPs and PCPs are in place (and they are). But what would happen if the food container the employees stored the chicken salad in was not properly cleaned and sanitized, dried, and stored away from dirty dishware (like pans used for raw chicken) prior to its use? If a cleaning and sanitation FSMS was not designed and implemented, the risk of this happening could be high and lead to many customers getting sick from this chicken salad as it is portioned and served over several days. Thus, these additional FSMS based on the Prerequisite Control Program are critical to the prevention of foodborne illness and outbreaks.

Remember that a FSMS as defined by the FDA should include procedures (SOPs), training (what and how), and monitoring by a knowledgeable person that can make corrective actions based on the proper procedure. It is not in the scope of this book to describe each of the Prerequisite Control Program components that which could require a separate FSMS. This is because each component may be more or less advanced and managed differently by the foodservice business based on its size and number of locations. Thus several of the components may be combined into a single FSMS. For example, pest prevention, facilities and equipment maintenance, safe chemical use, allergen prevention, and cleaning and sanitation could all be developed into one FSMS. Likewise, all of the personal hygiene components of the Prerequisite Control Program (health policy, exclusion/restriction of sick employees, glove use, hand washing, etc.) could all be developed into one FSMS (in this example, some of these components would also need to be monitored in the Process HACCP FSMS at points of food preparation and service as PCPs; see above and Chap. 3). Regardless, you would need to define the procedures for each, the training for them, and the means to monitor them (e.g., checking to ensure each employee has been trained on personal hygiene and the health policy and performing an employee wellness check at each shift) to ensure all hazards associated with all the components of the Prerequisite Control Program are controlled. In order to show how important this will be in the prevention of foodborne illness and outbreaks, let me demonstrate a few examples where an FSMS is developed from a component of the Prerequisite Control Program.

## *Cleaning and Sanitation Food Safety Management System*

One of the most important FSMS in a foodservice establishment that will directly impact the efficacy of the Process HACCP based FSMS is the cleaning and sanitation management system, which should include:

1. The proper chemistry that will clean and sanitize surfaces to remove hazards (effective detergents/cleaners) and kill pathogens quickly (effective no-rinse sanitizers) like Norovirus or Hepatitis A

2. The proper tools to effectively clean foodservice environments that don't add to cross-contamination risk. Separation of cleaning tools based on task—restrooms vs. kitchen
3. Effective procedures that can be executed properly by employees even if not trained
4. Trained employees who are empowered to perform the procedures correctly and at the proper times
5. A monitoring/corrective action-based process to prevent cross-contamination risk during foodservice operations

As discussed above, one of the top five risk factors for the cause of foodborne disease illnesses in restaurants is due to cross-contamination of food from unclean/unsanitary food utensils, storage containers, and food contact surfaces. These cross-contamination events could likely be reduced with proper cleaning and sanitizing of food contact and non-food contact surfaces, proper dish washing, and sanitation of high-touch point (described below) environmental surfaces using the same method of identifying the hazards, where and when they are most probable to occur in the facility, and then using the FSMS to monitor to ensure compliance to the SOPs.

The purpose of foodservice cleaning and sanitation SOPs, like hand washing and hard surface cleaning/sanitation, is to reduce the load/number of pathogenic micro-organisms on hands and surfaces that can make humans sick. Reduction of these pathogens is done with the intent to limit the amount that might cross-contaminate foods to a level that if there is consumption by someone, it will be below the infectious dose, in other words at a level low enough for a typical immune system to fight them off. As hygiene interventions have evolved, some have been found to be too risky for continuation. For example, bar soaps and open-refillable bulk soap systems have been shown to harbor pathogens and cause outbreaks—thankfully they are no longer allowed by CDC in US healthcare (Morbidity and Mortality Weekly Report 2002) and should not be used in foodservice facilities. Without a defined FSMS for cleaning and sanitation, there is also the risk for misuse of the chemicals and tools used in the facility that then actually contribute to the risk of cross-contamination and a resultant foodborne illness or outbreak.

One example of this need to develop a cleaning and sanitation FSMS to ensure all hazards are under control (that can also serve as one of the proper cleaning and sanitation SOPs in the FSMS) is in how the business uses and stores reusable towels for cleaning and sanitizing surfaces (summarized in King (2018)). The 2017 FDA Food Code allows for the use of reusable towels/cloths to wipe food contact surfaces and equipment, but the towels/cloths must be held between use in a chemical sanitizer solution at a concentration specified under Section 4-501.114 (FDA 2017). Presumably, storing the towels in a sanitizer solution should destroy the organisms picked up during the cleaning process. However, organic material present in sanitizing solutions can potentially bind to the active agent lowering the concentration below that which is effective. This allows organisms that are picked up during the wiping process to survive that treatment and be transferred to subsequent surfaces. Additionally, during the cleaning/sanitizing process, the towel being used may collect materials that would offer protection to the microbes in question. Fats and proteins may form films on the towels which can sequester microorganisms allowing them to "hide" from the active ingredients which are present in the sanitizer. These films may also be transferred to subsequent surfaces during the cleaning protocols, thus transferring their filth as well as microorganisms.

We also know that viruses can easily be transferred from contaminated surfaces onto previously clean surfaces by the reusable towels used to clean and sanitize surfaces (Gibson et al. 2012). One study found that typical cotton bar towels used by restaurants can remove approximately 3-log PFU (plaque-forming units, one method used to enumerate live viruses), but will transfer as much 2-log PFU back to surfaces (Gibson et al. 2012). Surveys of used cloths in food retail have been shown to be heavily contaminated with various bacterial species with one study from the published literature showing that 74% of cloths (n = 131) used for cleaning in raw and prepared food spaces were contaminated with *E. coli*, *S. aureus*, *E. faecalis*, and/or *C. perfringens* (Tebbutt 1986). Specifically, *E. coli* was isolated from 74 cloths with 25 of those carrying more than $10^5$ CFU (colony-forming units, one method used to enumerate live bacteria). There is also evidence to suggest that this gets worse the longer that the cloth towel is used (Tebbutt 1988). Single-use products (e.g. paper towels, single-use pre-moistened wipes, etc.) avoid these issues by being thrown away in the garbage after each use.

Likewise, if the employee is not trained properly on the proper SOP, they may only use the damp reusable towel to scrape food off surfaces assuming this action is cleaning and sanitizing, where a cleaner and a fresh sanitizer solution should be used. How often is this likely to be improperly used in a foodservice business; a recent survey of the restaurant industry (representing over 40 individual restaurant chains) and regulatory officials showed it is a significant concern (Fig. 4.2).

As an example of how important a defined SOP in a cleaning and sanitation FSMS can be, the proper SOP for using reusable towels (based on the 2107 FDA Food Code), summarized here, would include:

1. Store clean damp reusable towels/cloths in a container with appropriate sanitizer at the required concentration when not in use:

   (i) All wet reusable towels/cloths should be laundered (cleaned) or discarded daily. The wiping cloths may be laundered in a mechanical washer, a sink designated only for laundering wiping cloths, or a ware washing equipment (dishwasher) or food preparation sink that is cleaned and sanitized before use.

   (ii) All wet reusable towels/cloths used for wiping surfaces in contact with raw animal foods should be stored and used separately from wet reusable towels/cloths used for other purposes (e.g., ready-to-eat (RTE) food prep surface).

   (iii) All wet reusable towels/cloths and the sanitizer solution in which they are held between uses should be free from food debris and visible soil.

   (iv) The containers of chemical sanitizer solutions used to store wet reusable towels/cloths between uses should be stored off the floor and used in a manner that prevents contamination of food, equipment, utensils, linens, and single-service/use articles.

2. Clean and sanitize food contact surfaces with reusable towel/clot:

   (i) When cleaning a food contact surface, remove the wet towel from solution, and squeegee excess solution off the towel/cloth.
   (ii) Scrape off any visible food debris off the surface.
   (iii) Clean the surface properly to remove organic material including oil, grease, fat, and remaining food debris by applying an appropriate detergent (e.g., spray or pour) on the surface and wiping the surface clean to sight and touch.
   (iv) Apply a chemical sanitizer of adequate temperature and chemical concentration and allow it to remain on the surface for the specific contact time (according to the manufacturer's Environmental Protection Agency (EPA)-registered label).
   (v) Allow chemical sanitizer to air-dry on the surface or wipe the sanitizer off the surface using the reusable towel/cloth.

3. Store clean damp reusable towels/cloths in a container with appropriate sanitizer at the required concentration when not in use.

The call to action from this information is to of course use and develop all the proper procedures SOP's like those defined above into the cleaning and sanitation FSMS (for safely using reusable towels/cloths to clean and sanitize food contact surfaces). However, as discussed above (and see Fig. 4.2), it may be difficult to gain full compliance to this cleaning and sanitation SOP (e.g., identified during an assessment like a third party audit) which then increases the risk of foodborne illness and outbreaks. Thus, using alternative methods/tools in the FSMS may be important to remove this risk and increase compliance to proper cleaning and sanitizing of food contact surfaces. These alternative methods/tools could include:

* Single-day use (used for 1 day and discarded) disposable wiping towels/cloths that do not absorb all current sanitizers in use
* Reusable towels/cloths that prohibit biofilms/grease/oil/fats and pathogen survival on surface (antimicrobials built in to the towel/cloth)
* Validated towel/cloth laundering methods that eliminate biofilms/grease/oil/fats and kill pathogens
* Food contact surface sanitizers (used to store towels/cloths) that kill all foodborne disease-causing pathogens including (but not inclusive of all) Norovirus, *Clostridium perfringens*, *Listeria*, Hepatitis A, *Campylobacter*, and *Salmonella*
* Sanitizer solution concentration indicator technology or better methods to alert user to when the sanitizer concentration has dropped below required levels
* Single-use disposable wipes that kill all foodborne disease-causing pathogens for use on high-touch point surfaces and lightly soiled food contact surfaces as replacement of reusable towels/cloths (it is likely that safe use of reusable tow-

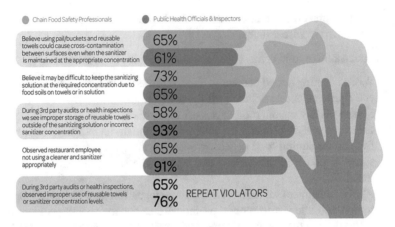

**Fig. 4.2** In order to assess the important opinions of different foodservice food safety professionals (those who lead food safety in their respective corporate businesses including QSR, fast-causal, casual-dining, and cafeteria/buffet restaurants) and local regulatory authority professionals (those who audit and enforce compliance to their state food code requirements in three states including Ohio, Texas, and Virginia) for their perception of the risk associated with use of reusable towels in foodservice, a survey was developed (in the months of May to July 2018). The survey was designed for each, respectively, and open only to confirmed participants based on solicitation to their respective organization via a collaboration between Public Health Innovations LLC and the National Restaurant Association. For more detailed information, see https://www.thefoodsafetylab.com/blogs/is-it-time-to-change-how-we-clean-and-sanitize. (Source: © 2018 GOJO Industries Inc.)

    els/cloths would still be needed for cleaning heavily soiled equipment and other surfaces in the foodservice establishment)

- Improved color-coded storage of reusable towels/cloths for cleaning raw animal food prep surfaces (e.g., yellow container and towels/cloths for raw chicken) and RTE food prep surfaces (white container and towels/cloths)

    The selection of the cleaner and sanitizer is critical to the effective cleaning and sanitation—based FSMS. The cleaner (detergent) should easily remove oils, fats, and grease and enable biofilm removal with scrubbers and brushes used as cleaning tools. The sanitizer should have a long shelf life when not in use, be active against the majority of the foodborne pathogens/hazards likely to be present in the foodservice facility (Norovirus, Hepatitis A, *Salmonella*, *E. coli*, *Staphylococcus*, *Campylobacter*, *Clostridium*, etc.), and have efficacy within 60 seconds or less contact time on cleaned surfaces. Preferably the sanitizer should be no-rinse to reduce the risk of re-contamination of a surface with unclean or soiled water.

    Chlorine and quaternary ammonium compounds (QACs) are two common active ingredients used for no-rinse, food contact surface sanitation in foodservice environments. Both are known to be less effective in the presence of hard water or organic soils and especially when targeting bacteria dried onto surfaces (Best et al. 1990; Shen et al. 2013; Jono et al. 1986). For example, the addition of just 1% milk

e.g., transmission occurs via employee contaminating hands while using a restroom then touches surfaces in kitchen

e.g., other employees touch these same handles and then prep RTE foods (even with gloves on)

**Fig. 4.3** The common risk of biological hazard (e.g., Norovirus) transmission from the restroom environment (restroom door knobs, restroom sink faucet handles/knobs, restroom stall latches, etc.) to the kitchen surfaces (oven door, refrigerator door, cook equipment buttons, etc.) that can then lead to cross-contamination of RTE foods by hands or gloved hands. (Source: King and Michaels 2019)

(semi-skimmed) has been shown to dramatically reduce the activity of QAC solutions (Lambert and Johnston 2001). In addition, the EPA testing for food contact surface (FCS) sanitizers does not require additional soil load to be added to the surfaces being sanitized; thus, approved sanitizers should only ever be used on clean surfaces (AOAC 2013); use a cleaner on the surface first, wipe it off, and then sanitize the surface. However, care does need to be taken to ensure that the cleaning agent used does not interfere with the sanitizers. For example, QACs may be inactivated by binding with anionic surfactant-based soap residues.

Another important need in the cleaning and sanitation FSMS is a focus on transmission of biological hazards onto non-food contact surfaces. One of the most common biological hazards and the one that causes the largest number of foodborne disease outbreaks in the United States (CDC 2018) that must be prevented by multiple controls is Norovirus. The Process HACCP-based FSMS has as its focus the prevention of the contamination of foods from working sick employees via a wellness check (to try and restrict employees who may have Norovirus but also Hepatitis A and other foodborne illnesses including *Salmonella* infection) and the monitoring for the proper use of hand washing and glove use as barriers to protect food from employees' hands. However, an additional risk from the hazard of Norovirus or other foodborne pathogens spread by sick employees is in the transmission of viruses to non-food contact environmental surfaces within the restaurant.

The contamination of hands in the restrooms by food handlers, directly from the source or from objects in the restroom, is the major route of Norovirus transmission to the retail foodservice environment (Fig. 4.3). If a symptomatic food handler (an employee experiencing diarrhea and/or vomiting due to Norovirus) is not excluded or not detected via a requirement of employees to report these symptoms, a very high number of customer infections will occur during an outbreak (Duret et al.

2017). The high levels of infected customers when a food handler works while sick (diarrhea and/or vomiting) are explained by the high level of Norovirus introduced into the foodservice environment due to the employee's frequent visits to the restroom.

When symptomatic and asymptomatic food handlers work while sick or additional conditions change that increase the likelihood of Norovirus entering the foodservice environment in higher numbers (e.g., community-associated outbreak in the local high school or customer/employee vomiting in the restaurant), a significant increase in the number of expected customer infections can be affected by the presence of the virus on high-touch point surfaces in the restroom and kitchen areas. A cleaning and sanitation FSMS is even more important to prevent a foodborne disease outbreak. This risk can be mitigated via the FSMS sanitation control of these high-touch point surfaces using a sanitizer that kills Norovirus quickly upon contact (spray and leave on surface) and escalating the times and additional surfaces to sanitize when the risk escalates due to these events.

A foodservice sanitation "Hot Spot Map" program was designed (based on an FDA-led risk model on the transmission of Norovirus in foodservice establishments) in a collaboration between Public Health Innovations and GOJO/Purell to assist the restaurant industry in this practice of using a FSMS, by showing which surfaces are high-touch point surfaces (i.e., where the hazards are most likely) to lead to transmission of Norovirus and when to perform the Norovirus sanitation of these high-touch point surfaces (see https://www.foodsafetymagazine.com/products/Norovirus-hot-spot-program-for-restaurants-launched-by-purell-and-public-health-innovations/).

Based on the scientific review of Norovirus transmission risk in the foodservice environment, it was recommended that the restaurants (based on key employee and customer touch points that are most likely to occur in a foodservice facility in Fig. 4.4) perform:

1. **Routine prevention**—Sanitize each surface area (noted by indicated dots) once per shift when:

   (a) NO employee has reported being sick nor has been excluded from work due to illness that includes diarrhea and/or vomiting.
   (b) There are NO reports of a community-associated Norovirus outbreak (e.g., in local high schools).
   (c) There have been NO body fluid spills by an employee or customer (e.g., vomiting anywhere in the restaurant).

2. **Escalated prevention**—Sanitize all restroom surfaces (e.g., restrooms) every hour and continue sanitation of all other surfaces each shift when:

   (a) An employee has reported being sick or has been excluded from work due to illness that includes diarrhea and/or vomiting.
   (b) There are reports of a community-associated Norovirus outbreak (e.g., in local high schools).

## 🕍 NOROVIRUS HOT SPOT MAP™: Key Touchpoints

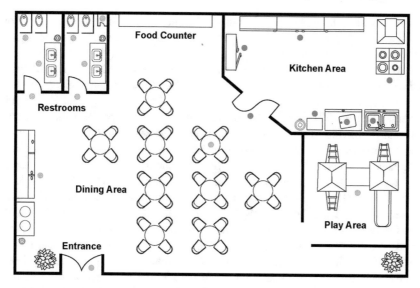

**Fig. 4.4** Example of control for Norovirus transmission to facility environmental surfaces in a cleaning and sanitation FSMS. The highest risk of transmission of Norovirus occurs in the restroom environment by a sick employee (dots at restroom door knobs, restroom sink faucet handles/knobs, restroom stall latches, etc.) and then leads to transmission to kitchen surfaces (dots at oven door, refrigerator door, cook equipment buttons, etc.) by this and other employees that can then lead to cross-contamination of RTE foods by hands or gloved hands; see Fig. 4.3. (Source: © 2019 GOJO Industries Inc.)

    (c) There has been NO body fluid spill by an employee or customer (e.g., vomiting anywhere in the restaurant).

3. **Maximum prevention**—Sanitize all surfaces every hour when:

    (a) An employee has reported being sick or has been excluded from work due to illness that includes diarrhea and/or vomiting.
    (b) There have been reports of a community-associated Norovirus outbreak (e.g., in local high schools).
    (c) There has been a body fluid spill by an employee or customer (e.g., vomiting anywhere in the restaurant).

Norovirus outbreaks in a community where a restaurant operates can increase the risk of Norovirus (and likely Hepatitis A) contaminating surfaces in a restaurant from the following:

- Many high school students also work in restaurants (many of whom may be asymptomatic).
- There may be more vomiting and diarrhea events in restaurants from employees and customers.

- More customers may carry Norovirus, and transmit the virus to common use surfaces.

Thus, ensuring the FSMS for cleaning and sanitation is focused on the foodborne illness hazards and not just cleaning the restaurant to look clean must be a part of the restaurant's food safety management program.

An important risk when using a cleaning and sanitation FSMS is in the actual use of chemicals during food storage and preparation. Chemicals especially any pesticides (which should only be used by licensed professionals (an meet EPA requirements—USEPA 2018), but many restaurants attempt to perform their own pest control via buying over-the-counter pesticides; see below) or disinfectants must be stored and used away from all food prep. Pest control is best left to certified vendors that have provided effective pest control to restaurant facilities (not the "spray and go" type), but you must ensure the cleaning and sanitation FSMS is in place to prevent the contributing factors that lead to pest infestations such as dirty reusable towels with food soils left for days, food crumbs and dust (e.g., flour) on floors, open food in dry storage areas, standing water on floors, etc. You should partner with a foodservice pest control program provider, one that has demonstrated the methods to "proof" facilities (see Chap. 6) against rodents and uses pesticides appropriate for foodservice establishments, and pesticides should not be stored in the restaurant, period. There are also additional risk to employees and associated fines from the Occupational Safety and Health Association (OSHA) if a restaurant business does not train employees on proper use of all cleaning and sanitation chemicals, cautions them on the hazards of use, and provides PPE (personal protection equipment) for their use where required according to the SDS (safety data sheet) program. Because this book is focused on the foodborne illness prevention, the reader can learn more about employee chemical safety and requirements at the OSHA website (see https://www.osha.gov/SLTC/restaurant/index.html).

## *Emergency Food Safety Management System*

An emergency FSMS is another FSMS critical to ensure preparedness for food safety-related emergencies that will occur in a restaurant but do not occur daily (nor are common), but must have a defined SOP, training, and monitoring to ensure employees are ready to act when the emergency does occur. These well-known emergencies occur every day in a foodservice business somewhere in the United States, and you should have written procedures that will be followed by CFPMs in the restaurant that include:

- **Body fluid clean-up** procedures and tools—Prescribing chemicals and tools available that will disinfect viruses like Norovirus, protect employees from exposure (PPE), and prevent transmission of the viruses to other restaurant surfaces. This is also an escalation event (see above) where you should apply the Norovirus

"Hot Spot Map" program to increase cleaning and sanitation against Norovirus in the restaurant environment (see above).

- **Sick/allergic customer calls/notices**—Prescribing how you will respond to a customer complaint/claim of a foodborne illness (see Chap. 8 for example of SOP for responses)
- **Power outages**—Prescribing how to prepare for and take actions when the power is out and refrigerated/frozen foods may be out of temperature compliance for more than 4 h
- **Boil water/outage notices**—Prescribing when it is possible to operate the restaurant safely when potable water has been lost including prior approval with the local health department
- **Pest infestation**—Prescribing how to operate the restaurant when there is an active infestation of roaches or rodents, for example, and when you should close the restaurant
- **Natural disasters**—Prescribing how to prepare for situations like a hurricane or other disasters that may lead to flooding, loss of power, etc.

The best resource for restaurants to develop SOPs for each of these is the most current recommendations made by the Conference for Food Protection, *Emergency Action Plan for Retail Food Establishments* (CFP 2014), based on the FDA Food Code specifications. Most states will also require a foodservice business to submit a plan for action/SOP for most emergency operations that would allow the business to remain open during/after some of these emergencies (see below). This document provides the most useful recommended SOPs (and those that most local and state health departments will support) on how a foodservice business may remain open and should operate during interruption of electrical service and water, when there is contaminated water including sewage backup, and after flooding and fires. This covers the majority of the emergency situations a restaurant business will experience that could impact food safety and restaurant operations but, when under the proper controls, allow the restaurant business to remain open, a considerable cost if not. Other SOPs not covered in this document that may be useful (there are other non-digital resources as well) are available as mobile apps as well, such as EmergiProtect (see https://www.activefoodsafety.com/emergiprotect). Digital food safety technology (see Chap. 7) like a mobile app that is based on credible SOPs like the Conference for Food Protection's *Emergency Action Plan for Retail Food Establishments* (CFP 2014) can be very helpful at the time of an emergency, as they are updated in real time and enable the manager to take the appropriate action quickly especially when the risk to human health is imminent (e.g., an allergic reaction to a food). If you choose to use paper-based SOPs, ensure they are always updated and reflect the information in this CFP document (and watch for updates to this document in the near future), and ensure your employees are trained on how to execute them.

You should also be aware of the FDA Food Code (see Section 8-404.11 and 8-404.12, FDA 2017) requirements that most state and local health departments will expect a restaurant to follow during an imminent health hazard to customers. The FDA considers an imminent health hazard as anytime there is a fire, flood, extended interruption of electrical or water service, sewage backup, misuse of poisonous or toxic materials, onset of an apparent foodborne illness outbreak, gross unsanitary occurrence or condition, or other circumstances that may endanger public health. The FDA states that a foodservice business shall immediately discontinue operations and notify the local health regulatory authority if any of these conditions exists. Considering the nature of the potential hazard involved and the complexity of the corrective action needed, the health regulatory authority may agree to continuing operations in the event of an extended interruption of electrical or water service if:

- A written emergency operating plan has been approved.
- Immediate corrective action is taken to eliminate, prevent, or control any food safety risk and imminent health hazard associated with the electrical or water service interruption.
- The health regulatory authority is informed upon implementation of the written emergency operating plan.

If operations are discontinued as specified under § 8-404.11 or otherwise according to law, the permit holder shall obtain approval from the local health regulatory authority before resuming operations and business.

## Supply Chain Food Safety Management System

One of the most important functions of FSMS based on Process HACCP (see above) is to verify that the food being used for the preparation of food products is from safe sources and has not, for example, been recalled by the FDA/USDA due to association with a foodborne disease outbreak or contains undeclared allergens. The best means to confirm food ingredients and products are from safe sources is to ensure each is from an FDA and/or USDA-regulated and inspected food manufacturing business, and each of its facilities is compliant (for complete details on how to manage a supply chain FSMS for a restaurant business with multiple locations, see King and Bedale (2017)).

All human food manufacturers (any ingredient, produce, or product) must now be in compliance to the Food Safety and Modernization Act (FSMA; see https://www.fda.gov/food/guidance-regulation-food-and-dietary-supplements/food-safety-modernization-act-fsma). FSMA, enforced by FDA inspections, requires all food manufacturing facilities (not just the brand name corporate business but their actual facilities) to define all the hazards associated with a product (ingredient and facility related), the exact preventive control for each hazard, and how they will monitor to ensure each control is functional during each product production run (similar to Process HACCP). The facility must also document this including the monitoring results via a **Food Safety Plan** (Fig. 4.5) for each product. All of your food ingredient and product suppliers should have evidence of compliance to this FDA required **Food Safety Plan** for any ingredient/product that they make and sell to your foodservice business. The FDA requires this for each human food manufac-

*Cookies Are Good for You? Inc.*

**Food Safety Plan**
**for**
**Burger Circus Cookie Dough**

Saint Simons Island, Georgia Facility

Reviewed by: ____Hal King____    Plant Manager  Date: ____10-06-2019____

**Table of Contents**

Company Overview

*[Provide a brief description of the company. Consider listing members of the food safety team, if you have one.]*

Product Description

| Product Name(s) | Raw Cookie Dough- CC |
|---|---|
| Product Description, including Important Food Safety Characteristics | Product made as RTE food for use in milkshakes, deserts, and candy |
| Ingredients | Milk, flour, eggs, chocolate chips, vegetable oil, salt, sugar |
| Packaging Used | 1 lb containers- 4 per case |
| Intended Use | RTE |
| Intended Consumers | Foodservice B2B only as ingredient for products sold to customers- not for individual sale |
| Shelf Life | 6 months unopened- 30 days upon opening with refrigeration |
| Labeling Instructions | Include allergen statement for milk, wheat, and eggs |
| Storage and Distribution | Refrigerated – 41°F all times |

| Approved: Signature: | Date: 2-7-2020 |
|---|---|
| Print name:   John Makeitsafe | |

**Fig. 4.5** Example Food Safety Plan document (TOC and product description on first two pages shown for example) that a product supplier is required to develop for each human food product manufactured under the FSMA rules enforced by the FDA. This plan should be available upon request for any product sold to a foodservice business

turing facility regardless of size, and the supplier should not have any reason not to share this information with you if they want your business. For more comprehensive details including "how to" guides, see *Hazard Analysis and Risk-Based Preventive Controls: Improving Food Safety in Human Food Manufacturing for Food Businesses* (King and Bedale 2017). The USDA has similar requirements for raw animal products and RTE meat, poultry, and fish products.

A second but equal FSMS function to ensure hazards are controlled by suppliers (including supplier controls you defined in your Process HACCP plan) is to collect data that can be used to verify that each manufacturer's control system (e.g., HACCP if for seafood, dairy, or juices and FSMA's HARPC for all other foods) is being continuously executed as designed, corrective actions are being made at the time of verification (enabling hazard prevention maintenance), and information collected is being used to continuously improve the hazard prevention; this must all be documented (as the FDA requires it) and available for review. This may also be done by asking the supplier for any third-party audit data or GFSI- or ISO-based certification audit data of their facility. If you are purchasing ingredients and food products via a third-party distributor such as a broad-line food delivery company, they should be able to provide you with this information as well.

You must also ensure you are NOT using previously approved sources that have now become unsafe due to an FDA recall or other reasons, such as if the CDC and FDA are investigating a food ingredient or product and have issued a "do not consume" caution to consumers and/or to the foodservice and retail sales (grocery, c-store) industry. This may also be communicated by the state where the business is located. This has happened multiple times due to outbreaks related to the contamination of produce where there was no FDA recall but the CDC and FDA announced to the public not to consume certain produce products based on where the produce was grown or which supplier processed it, sometimes specific to the source, but more often only a state or region of where the produce was grown. You can find the most current recall of foods information and when CDC and FDA are investigating a food item here: https://www.fda.gov/food/recalls-outbreaks-emergencies/recalls-foods-dietary-supplements. You can sign up to receive official email alerts of FDA food recalls here: https://www.fda.gov/safety/recalls-market-withdrawals-safety-alerts. It is important to monitor these regularly to enable you to communicate to all foodservice locations in your business to remove the unsafe food from all food preparation and service.

Remember also that not all product defects will be discovered by the FDA, CDC, or USDA; a foodservice business should also be prepared to remove ingredients/products from food preparation and service when they discover the defect (**biological**, e.g., food smells or tastes spoiled or has mold, etc.; **chemical**, e.g., food has undeclared allergen and is not labeled properly; or **physical**, e.g., food has glass pieces in it) during any of the food preparation processes or if you are alerted to this by a customer. Calls from customers and the local health department may be the first report to you that there is a product defect, and they should not be ignored. Removal of food ingredients and products previously considered safe (approved) but found unsafe is an important control most businesses do not prepare for. However a foodservice business must be prepared and should have a FSMS to manage this risk factor (defined by the Prerequisite Control Program (see above) and described in more detail as part of a business's food safety management program in King (2013)). If food ingredients/products are only sourced from a broad-line distributor (there are many), then ensure this business has a process of notifications to you and the foodservice locations that buy and receive their food.

## Corporate Management of Food Safety Management System(s) in the Foodservice Locations

I have purposely not discussed the business management of FSMS functions of the corporate food safety program (e.g., that which would be found in a multi-location restaurant chain) because most of this information has been discussed previously in the first book of on food safety management (King 2013). Likewise, restaurants that are part of a multi-location restaurant chain have been shown to have fewer food safety compliance violations per health inspection compared to single-location businesses (Leinwand et al. 2017) as they are more likely to have a food safety professional that helps the business manage food safety. This was observed in the FDA restaurant risk study defining the degree of FSMS being used in the restaurant industry discussed in Chap. 3 and in more detail above (FDA 2018). However, it is important to discuss that it is expected that a food safety management program exists in a restaurant business (even if you own only one restaurant) led by a food safety management expert to ensure all control systems which include all systems/specifications (including food safety SOPs), training and education, and facility design are appropriate, based on the most current food safety science and regulatory requirements, and each can be executed (i.e., validated that they can be performed as designed). The food safety management program responsibilities should include a process to select suppliers and to monitor each food ingredient and product manufacturer's control systems to confirm that all systems are working together to prevent biological, chemical, and physical product defects. Execution and verification should be measured by both corporate and third-party evaluation and the data then used to continuously monitor the effectiveness of the hazard prevention expected at the supplier's facility.

# References

AOAC International (2013) AOAC official method 960.09 – germicidal and detergent sanitizing action of disinfectants. Official Methods of Analysis of AOAC International, Gaithersburg

Best M, Kennedy ME, Coates F (1990) Efficacy of a variety of disinfectants against *Listeria* spp. Appl Environ Microbiol 56(2):377–380

Centers for Disease Control and Prevention (2018) Norovirus. Preventing Norovirus infection. Available at:https://www.cdc.gov/Norovirus/preventing-infection.html

Conference for Food Protection (2014) Emergency action plan for retail food establishments, 2nd edn. See: http://www.foodprotect.org/media/guide/Emergency%20Action%20Plan%20 for%20Retail%20food%20Est.pdf

Duret et al (2017) Quantitative risk assessment of Norovirus transmission in food establishments: evaluating the impact of intervention strategies and food employee behavior on the risk associated with Norovirus in foods. Risk Anal 37(11):1–27. An FDA risk assessment

Food and Drug Administration (FDA) (2017) FDA food code. Recommendations of the United States Public Health Service, Food and Drug Administration, National Technical Information Service Publication number IFS17. Food and Drug Administration

Food and Drug Administration (FDA) (2018) FDA report on the occurrence of foodborne illness risk factors in fast food and full-service restaurants, 2013–2014

Gibson KE, Crandall PG, Ricke SC (2012) Removal and transfer of viruses on food contact surfaces by cleaning cloths. Appl Environ Microbiol 78(9):3037–3044

Jono K, Takayama T, Kuno M, Higashide E (1986) Effect of alkyl chain length of benzalkonium chloride on the bactericidal activity and binding to organic materials. Chem Pharm Bull 34(10):4215–4224

King H (2013) Food safety management: implementing a food safety program in a food retail business. Springer, New York

King H (2018) Is it time to change how we clean and sanitize food contact surfaces with reusable wiping towels?Food Safety Magazine. June/July. Available at:https://www.foodsafetymagazine.com/magazine-archive1/augustseptember-2018/is-it-time-to-change-how-we-clean-and-sanitize-food-contact-surfaces-with-reusable-wiping-towels/

King H, Bedale W (2017) Hazard analysis and risk-based preventive controls: improving food safety in human food manufacturing for food businesses. Elsevier, London

King H, MichaelsB (2019) The need for a glove-use management system in retail foodservice. Food Safety Magazine. June/July. Available at: https://www.foodsafetymagazine.com/magazine-archive1/junejuly-2019/the-need-for-a-glove-use-management-system-in-retail-foodservice/

Lambert RJW, Johnston MD (2001) The effect of interfering substances on the disinfection process: a mathematical model. J Appl Microbiol 91(3):548–555

Leinwand SE, Glanz K, Keenan BT, Branas CC (2017) Inspection frequency, sociodemographic factors, and food safety violations in chain and 77 nonchain restaurants, Philadelphia, Pennsylvania, 2013-2014. Public Health Rep 10:1–8

Luning PA, Bango L, Kussaga J, Rovira J, Marcelis WJ (2008) Comprehensive analysis and differentiated assessment of food safety control systems: a diagnostic instrument. Trends Food Sci Technol 19(10):522–534

Luning PA, Marcelis WJ, Rovira J, Van der Spiegal M, Uyttendaela M, Jacxsens L (2009) Systematic assessment of core assurance activities in a company-specific food safety management system. Trends Food Sci Technol 20(6):300–312

Morbidity and Mortality Weekly Report (2002) Recommendations and reports. 51(RR-16)

Shen C, Luo Y, Nou X, Wang Q, Millner P (2013) Dynamic effects of free chlorine concentration, organic load, and exposure time on the inactivation of *Salmonella, Escherichia coli* O157: H7, and non-O157 Shiga toxin–producing *E. coli*. J Food Prot 76(3):386–393

Tebbutt GM (1986) An evaluation of various working practices in shops selling raw and cooked meats. EpidemiolInfect 97(1):81–90

Tebbutt GM (1988) Laboratory evaluation of disposable and re-usable disinfectant cloths for cleaning food contact surfaces. EpidemiolInfect 101(2):367–375

U.S. Environmental Protection Agency (2018) Pesticides, tolerance exemptions for active and inert ingredients for use in antimicrobial formulations (food-contact surface sanitizing solutions). Fed Reg 69:23113–23142. Available at:https://www.ecfr.gov/cgi-bin/text-idx?SID=fb2d71d9f2edfd0d6cdba8471006bb55&mc=true&node=se40.26.180_1940&rgn=div8

# Chapter 5
# Training to Enable Food Safety Management Systems

Employee training and knowledge are critical to laying the foundation for the Food Safety Management Systems (FSMS) necessary to enable Active Managerial Control of food safety risk in a foodservice environment. First, training should include the advanced food safety management knowledge (e.g., menu/recipes, HACCP certification, SOP development, monitoring, and corrective actions) for owners/operators and general managers to enable them to design and implement the FSMS based on the menu (see Chaps. 3 and 4). All other employees' food safety training (manager/supervisor, kitchen manager, food handler, food server) should be based on what the FDA has prescribed as important to the management of food safety risk factors in the FDA Food Code (FDA 2017) called demonstration of knowledge. Training to this core knowledge is key to ensuring that each employee understands the top five foodborne illness risk factors and the necessary controls they will use in their role as food handler or manager (specific to the foodservice establishment type and menu served).

Employee knowledge of how hazards occur (e.g., Norovirus transmission from the restroom to the kitchen and then cross-contamination of food by hands) and how they are controlled (not working when sick, washing hands, and wearing gloves properly) is critical to the effective FSMS. This knowledge by all employees will also enable the reinforcement of food safety education as food handler employees practice the knowledge and as managers monitor the controls during food preparation processes, as employee education is also influenced by observing others.

Employee behaviors and their resultant actions play a significant role in the control of all hazards in the foodservice business. The CDC and others have defined 32 total contributing factors to foodborne disease outbreaks from foodservice establishments in the United States (Bryan et al. 1997; CDC 2014; Appendix A). These fall under three broad categories of procedural/management in food preparation:

- Contamination—situations in which pathogens and other hazards contaminate ready-to-eat food (15 contributing factors)

© Springer Nature Switzerland AG 2020
H. King, *Food Safety Management Systems*, Food Microbiology and Food Safety, https://doi.org/10.1007/978-3-030-44735-9_5

- Proliferation—situations in which pathogens grow and/or their toxins are produced in ready-to-eat food (12 contributing factors)
- Survival—situations in which pathogens survive a process designed to kill or reduce their numbers (five contributing factors)

In a recent report from the CDC's National Environmental Assessment Reporting System (NEARS) on foodborne disease outbreak characteristics, 81% of 114 outbreaks from foodservice establishments in 2015 involved a contamination event as a contributing factor (CDC 2017a). The NEARS program at CDC was developed (in part) to better understand the contributing factors of single-setting foodservice establishment outbreaks so that research could be performed on which interventions are most effective in preventing them (Brown et al. 2017; also see Chap. 2).

Interestingly, all of the 32 known contributing factors point to a lack of employees who follow proper procedures and/or a lack of a managerial oversight in taking actions to prevent their occurrence. Four of the top common contributing factors are directly related to employees not following correct procedures and managers not taking corrective actions during the preparing and handling of food (CDC 2017b), including situations in which:

- A sick food handler contaminates ready-to-eat food through bare-hand contact.
- A sick food handler contaminates food through a method other than bare-hand contact (e.g., using a contaminated utensil).
- A sick food handler contaminates ready-to-eat food through glove-hand contact (contaminated glove).
- Food handling practices lead to growth of pathogens (such as failure to maintain food at a sufficiently cold temperature).

One of the best examples of effective knowledge linked to employee behavior and management of food safety risk to prevent contributing factors is a written health policy to ensure that sick employees do not prepare and serve food (a health policy and its relationship to the FSMS are discussed in more detail in Chaps. 3 and 4). Briefly, a health policy requires a manager to know for which signs, symptoms, and diseases they should be vigilant, regularly educating, inquiring, and monitoring employees to ensure that sick food handlers are excluded/restricted from preparing and serving foods. The food handlers themselves must also be educated about which signs, symptoms, and diseases must be reported to a manager/supervisor—and also know not to prepare and serve food when they are sick. The training to this knowledge and roles is required education for each employee according to the FDA Food Code (i.e., food handlers should self-report when sick, and managers should screen and exclude/restrict employees properly). This is necessary in order for a health policy to be effective in reducing the risk of sick employees preparing food.

Food safety training for each employee should consist of a structured curriculum based on the prerequisite experience of the employee: (1) food safety requirements built into the food preparation task used to store, prepare, and serve all food (e.g., based on recipes), (2) a means to deliver the training to diverse students (e.g., differences in age, language, amount of foodservice experience), (3) hands-on demonstration, and (4) a method to document education of the student via knowledge test and visually observed and assessed application of the knowledge. Finally, the training curriculum should be offered in a delivery method that will accommodate the students' means to demonstrate competency (e.g., online course vs. lecture); food safety education (after the training) must be verified on a regular basis (audits and performance-based evaluations) to determine when additional training or re-training is necessary: for example, when training must be used as a corrective action; as with an employee who has been repeatedly observed not washing hands correctly (for more details on training as a corrective action, see Chap. 4).

## Training Necessary to Achieve Demonstration of Knowledge of Foodborne Illness Risk Factors

In addition to the owner/operator, a foodservice establishment operation has two primary positions for which employees must know the basic knowledge (food handlers) or both the basic and advanced levels of food safety knowledge (managers and kitchen supervisors/chefs). The owner/operator must know both, and the ability or hire/contract for the ability to train employees on both. All of the food safety training materials should be based upon a foundation of the most current knowledge concerning food safety risk in a foodservice business, not just the risk of an outbreak that may be rare but risk of a single foodborne illness which is common (see Chap. 2). The training must align with the most current FDA Food Code and include the specific procedures, tools, and monitoring methods necessary to prevent hazards in food on a daily basis. Aligning all food safety training curriculum to the most current FDA Food Code is critically important and repeated here because the controls used in the FSMS will be based on this knowledge.

Aligning all food safety training curriculum to the most current FDA Food Code is important to an independent foodservice establishment business operating in one state but also for numerous, similar foodservice concepts operating in different states as part of a foodservice chain or franchised business. This is because the majority of local and state food codes are based on a version of the FDA Food Code and are enforced via local inspections (for which health inspectors use risk-based inspection scoring/grading of the foodservice establishment to communicate risk to the public). Although local and state health departments may have established their food code rules based upon an older version of the FDA Food Code, it is still advantageous for foodservice establishments to align all training to the most current FDA Food Code. This will ensure the most current knowledge on food safety management is used when selecting the training curriculum. This is most easily accomplished by choosing ANSI-certified (American National Students Institute) courses (see below), each of which aligns to the most current FDA Food Code standard, for food handlers, managers, and owners/operators/general managers.

# Role of Owners/Operators/General Managers

If you own and/or supervise (e.g., as a general manager of a franchised restaurant) or select persons who will own/lead a foodservice business operation, you/they should have this knowledge (defined in the Prerequisite Control Program for the business (see Chaps.3 and 4)):

- The design of your foodservice facility and/or establishment of the flow of food for food safety (see Chap. 6)
- What food products you are selling your customer
- Where the food ingredients you use to make food products come from and how to ensure their safety
- How to make the food products you sell
- Who makes the food products and when
- Where the food products will be prepared and stored
- What your quality requirements will be for prepared food (date of expiration if less than 7 days) and how you use FIFO
- How you will clean and sanitize dishware, equipment, and food contact surfaces and ensure chemical safety
- What you will use to train employees on how to prepare and serve the food products
- Who will train the employees
- Development of the health policy and exclusion/restriction of sick/injured employees
- How you will ensure pest prevention in the facility
- Who and how you will manage food safety risk and who will serve as the person in charge (PIC) with related responsibilities at all times of operation
- How you will serve the customers and with what service items (disposable or re-usable)
- Where customers will eat the foods you prepare and serve
- Whether there will be any self-service foods or beverages and how you will keep them safe during service

The food establishment owner/operator is likely a trained business professional and/or a food chef/expert (e.g., a restaurant chain franchisee business owner). This role entails knowledge of the menu and recipes used to prepare and serve food products; in addition, the owner/operator likely selects the approved food suppliers. Moreover, this person should have the most knowledge about how to design FSMS and train employees on FSMS within the establishment. At a minimum, the owner/operator should have the same level of training as a Certified Food Protection Manager (CFPM; note that many states require that a foodservice business owner must be certified; see below) and training on the recipes and SOPs required to operate the business. However, the owner/operator should also have more advanced training with HACCP certification (especially if an independent operator/owner) to enable application of Process HACCP methods and the design of FSMS.

As discussed in Chap. 1, Process HACCP is more appropriate to foodservice establishments than HACCP designed for manufactured foods; however, there are currently very few specific training courses on Process HACCP (thus a focus of this book). Therefore, a foundation of HACCP certification will be important to the effective design of FSMS. The FDA has published a resource (Managing Food Safety: A Manual for the Voluntary Use of HACCP Principles for Operators of Foodservice and Retail Establishments, FDA 2006) to help owner/operators of foodservice establishments learn more about using Process HACCP. I recommend that an operator/owner attain HACCP certification, obtain and review the FDA's manual, and then utilize this book as a means to support development of the FSMS necessary to enable implementation of Process HACCP in their establishments. It would also be beneficial for the owner/operator to complete the CFPM trainer course so that he/she can then train and proctor exams for their managers to be CFPMs.

## Role of Managers

Although certification alone is not a significant factor in the prevention of food-borne disease outbreaks (i.e., unless applied via actual management of food safety risk factors), there is clear evidence that indicates that manager-level certification (as evidence for the impact of prior training) in food safety knowledge can be correlated to the following:

- Restaurants with a CFPM present during operations had fewer foodborne illness risk factors out-of-compliance than those without a CFPM when inspected by the FDA (FDA 2018).
- Increased manager food safety demonstration of knowledge (Brown et al. 2014).
- Safer restaurant food preparation practices (Brown et al. 2014).
- Better health department inspection scores (Cates et al. 2009).
- Fewer foodborne illness outbreaks (Hedberg et al. 2006).

Therefore, all managers of all foodservice establishments should be trained at the CFPM level of food safety education. This training can be obtained through an American National Standards Institute (ANSI) Food Safety certification standard course (e.g., by certification bodies like National Restaurant Association's ServSafe, Prometric's Certified Professional Food Manager (CPFM), or National Environmental Health Association's Certified Professional in Food Safety (CP-FS)). This curriculum standard—although they vary slightly in structure and delivery with each certification body—is aligned to the most current FDA Food Code and is accredited by ANSI based on this FDA standard. The training teaches a manager the principles of food safety hazard identification and the standard controls to prevent these known hazards.

In 2006, the CDC endorsed the presence of a CFPM in foodservice establishments (see Fig. 5.1), stating that having a CFPM on site to perform Active Managerial

**Fig. 5.1** Centers for Disease Control and Prevention graphic targeted to the foodservice industry recommending that managers be certified in food safety (Source: CDC 2016)

Control of food safety hazards during all operations of a foodservice establishment is one of the more important means to prevent a foodborne illness outbreak in a foodservice establishment (CDC 2006). This endorsement was based on a CDC study (Hedberg et al. 2006) showing that the most significant difference between restaurants that were not associated with foodborne disease outbreaks was the presence of a CFPM. This report and CDC's endorsement show that the education of the person in charge within a foodservice establishment has a direct influence on the behaviors of the employees handling foods and correlates directly to reduction of risk factors. More recently, the FDA Food Code (FDA 2017) has recommended that each person in charge (PIC) of a foodservice establishment should be a CFPM; some states mandate that a CFPM is the PIC on site during all operations of a food-service establishment.

The demonstration and use of the knowledge gained from certification is more important than merely holding a certificate that could be 3 years old, as oftentimes in foodservice businesses once managers complete the training, they are not exposed to the proper procedures and/or use of FSMS necessary to apply their knowledge to the foodservice business. In other words, they don't maximize the practice of food safety management as a CFPM without the guidance of an FSMS. In one study, investigators compared the specific knowledge of 254 CFPMs (Certified Food Protection Managers ability to demonstrate knowledge of the required food safety controls to prevent hazards that contribute to the foodborne illness risk factors) working in 211 foodservice establishments to critical violations documented on health inspection reports. These investigators found that knowledge gaps were significantly correlated with the same critical violations observed by health inspectors during food preparation (Burke et al. 2014). In fact, the FDA Food Code has long required that knowledge of critical foodborne disease prevention be demonstrated during foodservice operations by the person in charge and not just the demonstration of certification (as a CFSM certificate holder) as a means to show proficiency of the ability to manage food safety risk.

Because training and certification may have occurred several years prior to the current work being performed by the person in charge (some states allow for re-certification within 3–5 years), regular demonstration of risk management knowledge is critical to effective food safety management. A better means to measure this (the evidence of food safety knowledge by the person in charge) is described in the most current FDA Food Code (FDA 2017) reproduced here to show FDA regulatory requirement and *summarized* by me (as example) to highlight the role each has in the management of FSMS:

1. Describing the relationship between the prevention of foodborne disease and the personal hygiene of a FOOD EMPLOYEE: *The manager knows the hazards and their contributing factors caused by employees that must be controlled to prevent cross-contamination of food.*
2. Explaining the responsibility of the *Person in Charge* for preventing the transmission of foodborne disease by a FOOD EMPLOYEE who has a disease or medical condition that may cause foodborne disease: *The manager will monitor the controls to prevent food handler employees from working while sick and ensure that they are washing hands and wearing gloves properly.*
3. Describing the symptoms associated with the diseases that are transmissible through FOOD: *The manager can monitor for sick employees using a wellness check based on the health policy and exclude/restrict employees properly.*

4. Explaining the significance of the relationship between maintaining the TIME/ TEMPERATURE CONTROL FOR SAFETY FOOD and the prevention of foodborne illness: *The manager knows that all food must not be held at temperatures within the temperature danger zone of 41 °F to 135 °F for more than 4 h to prevent growth/toxin production by biological hazards such as bacteria.*

5. Explaining the HAZARDS involved in the consumption of raw or undercooked MEAT, POULTRY, EGGS, and FISH: *The manager knows which biological hazards are most associated with raw proteins.*

6. Stating the required FOOD temperatures and times for safe cooking of TIME/ TEMPERATURE CONTROL FOR SAFETY FOOD including MEAT, POULTRY, EGGS, and FISH: *The manager knows what temperature all foods should be cooked to as controls to biological hazards.*

7. Stating the required temperatures and times for the safe refrigerated storage, hot holding, cooling, and reheating of TIME/TEMPERATURE CONTROL FOR SAFETY FOOD: *The manager knows what temperature all foods should be held/stored at as controls for biological hazards.*

8. Describing the relationship between the prevention of foodborne illness and the management and control of the following:

   (a) Cross-contamination
   (b) Hand contact with READY-TO-EAT FOODS
   (c) Hand washing
   (d) Maintaining the FOOD ESTABLISHMENT in a clean condition and in good repair

   *The manager knows the Prerequisite Control Program requirements and can monitor these controls in FSMS.*

9. Describing FOODS identified as MAJOR FOOD ALLERGENS and the symptoms that a MAJOR FOOD ALLERGEN could cause in a sensitive individual who has an allergic reaction: *The manager knows which allergens are present in the foodservice facility and how they are controlled and will be monitored by the FSMS.*

10. Explaining the relationship between FOOD safety and providing EQUIPMENT that is:

    (a) Sufficient in number and capacity
    (b) Properly designed, constructed, located, installed, operated, maintained, and cleaned

    *The manager knows how to operate and maintain the equipment used to cook, cool, hold, and store foods.*

11. Explaining correct procedures for cleaning and SANITIZING UTENSILS and FOOD CONTACT SURFACES of EQUIPMENT: *The manager knows the Prerequisite Control Program requirements and can monitor these controls in FSMS.*

12. Identifying the source of water used and measures taken to ensure that it remains protected from contaminating sources such as backflow and

cross-connections: *The manager knows the Prerequisite Control Program requirements and can monitor these controls in FSMS.*

13. Identifying POISONOUS OR TOXIC MATERIALS in the FOOD ESTABLISHMENT and the procedures necessary to ensure that they are safely stored, dispensed, used, and disposed of according to LAW: *The manager knows the Prerequisite Control Program requirements and can monitor these controls in FSMS.*

14. Identifying CRITICAL CONTROL POINTS in the operation from purchasing through sale or service that, when not controlled, may contribute to the transmission of foodborne illness, and explaining steps taken to ensure that the points are controlled in accordance with the requirements of this Code: *The manager knows the Process HACCP plan and Prerequisite Control Program requirements, can monitor these controls in FSMS, and can make the appropriate corrective actions.*

15. Explaining the details of how the PERSON IN CHARGE and FOOD EMPLOYEES comply with the HACCP PLAN if a plan is required by the LAW, this Code, or an agreement between the REGULATORY AUTHORITY and the FOOD ESTABLISHMENT: *The manager knows the Process HACCP plan and Prerequisite Control Program requirements including those that are required by the local health authority, can monitor and document these controls in FSMS, and can make the appropriate corrective actions.*

16. Explaining the responsibilities, rights, and authorities assigned by this Code to the:

    (a) FOOD EMPLOYEE
    (b) CONDITIONAL EMPLOYEE
    (c) PERSON IN CHARGE
    (d) REGULATORY AUTHORITY

    *The manager as person in charge (PIC) can teach the employees and demonstrate to the regulatory authority how the FSMS functions to comply to the Food Code and prevent foodborne illness risk factors.*

17. Explaining how the PERSON IN CHARGE, FOOD EMPLOYEES, and CONDITIONAL EMPLOYEES comply with reporting responsibilities and EXCLUSION or RESTRICTION of FOOD EMPLOYEES: *The manager can describe the foodservice business's health policy and how it is used to reduce risk of working sick employees* (FDA 2010).

According to the 2017 FDA Food Code, additional knowledge and duties of the person in charge/food safety manager (CFPM) include a responsibility to ensure (rephrased without code references per foodservice use) the following (each easily managed with FSMS):

- Operations are not conducted in a private home or in a room used as living or sleeping quarters.
- Unnecessary persons are not allowed in food prep, food storage, nor dishware washing areas except for brief visits and tours if steps are taken to ensure exposed

food, clean equipment, utensils, linens, and unwrapped single-service and single-use articles are protected from contamination.
- Employees and other persons such as delivery, maintenance, or pest control persons must comply with food safety management requirements.
- Employees are effectively cleaning their hands by routinely monitoring employee hand washing.
- Employees are visibly observing foods as they are received to verify that they are received from approved sources, delivered at the required temperatures, protected from contamination, unadulterated, and accurately labeled.
- Employees are verifying foods delivered to the establishment during non-operating hours are from approved sources, and are placed into appropriate storage locations at required temperatures, protected from contamination, unadulterated, and accurately labeled.
- Employees are properly cooking food by monitoring the cooking temperatures.
- Employees are using proper methods to rapidly cool foods that are not held hot or not consumed within 4 h by monitoring food temperatures during cooling.
- Employees are properly maintaining the temperature of foods during hot and cold holding via daily monitoring of the temperature of foods.
- Consumers who order raw or partially cooked foods of animal origin are informed that the food is not cooked sufficiently to ensure safety.
- Employees are properly sanitizing cleaned multi-use equipment and utensils before they are used through routine monitoring of solution temperature and exposure time (for hot water sanitizing) and chemical concentration, pH, temperature, and exposure time (for chemical sanitizing).
- Consumers are notified that clean tableware is to be used when they return to self-service areas such as salad bars and buffets.
- Employees are preventing cross-contamination of food with bare hands by properly using utensils such as deli tissue, spatulas, tongs, single-use gloves, or dispensing equipment.
- Employees are properly trained in food safety, including food allergy awareness.
- Employees are informed in a verifiable manner of their responsibility to report to the person in charge information about their health and activities as they relate to diseases that are transmissible through food.
- Written procedures and plans are maintained and implemented throughout the establishment.

No current CFPM course certified by ANSI provides training on how to design a Process HACCP plan and Prerequisite Control Program, nor how to specifically execute a FSMS, but each is designed to train a manager on the FDA specific requirements for demonstration of knowledge, management responsibilities discussed above, and how to monitor controls of the hazards associated with the preparation of food. However, additional training and experience can be acquired by the managers in the use of FSMS designed by a Process HACCP plan and the Prerequisite Control Program, which can regularly reinforce the CFPM training.

## Role of Managers in the Monitoring and Assessment of Food Safety Management Systems

Managers must be provided the training and tools to determine how well the food-service establishments' current FSMS is working. Further, they must be enabled to make the appropriate corrective actions and determine root causes of new issues to prevent the same issue from reoccurring. Many of the corrective actions needed are specific to the foodservice establishment's processes and require manager-level decisions that can impact the safety of the food and the economics of the business. For example, if a raw animal food product does not cook properly in an open fryer of oil after the defined cook time, it should be recooked and then the next raw animal food product to be cooked on that same equipment rechecked to ensure the equipment is cooking properly. However, suppose that when the equipment is checked it is found to be working properly (e.g., oil temperature correct, cooking time correct, placement of raw animal food in cook basket correct) but the raw animal food that is received frozen is not thawed completely before it is being cooked, leaving the internal meat raw. Remember that the hazard here is that pathogens (like *Salmonella*) can survive the cooking process in partially frozen poultry, the contributing factor could be is the incomplete thawing of the chicken, and the foodborne illness risk factor is undercooked raw poultry likely not reaching 165 °F.

The root cause of this uncooked food could be due to many variables, each specific to the processes the establishment follows, including not using the correct time/temperature to thaw the product (an employee issue) or improper size/weight of the raw animal product specification to ensure product thaws under time allowance (a specification for the supplier). Uncorrected, either cause would likely lead to many undercooked products being served to customers that could have easily been avoided by monitoring the processes with manager-level corrective actions. Of course, the manager needs a pre-established well-defined Process HACCP plan to define what processes like this to monitor (as discussed in Chaps. 3 and 4), such as a requirement to check size/weight of raw animal foods, verification of thawing a food before cooking, and checking the temperature of the final cooked food.

In this example, in order for the manager to know which control to monitor that would have the best control of the contributing factor of undercooked chicken, he/she would need to have an SOP that defines the corrective action when the chicken does not cook properly: how to check size/weight of frozen chicken to determine if size/weight is impacting thawing, how to verify that chicken is thawed properly before cooking, and if not cooking to 165 °F, how long to recook the product or should it be discarded. For example, the owner/operator may require that all raw animal foods associated with undercooked raw animal food product assessments be recooked first and then the next batch of raw animal food checked again to ensure the cook equipment is working properly. The owner may include in this corrective action a requirement to check for thawing issues described above if the cook equipment is working properly but the product is still not reaching the required temperature and report any size/weight of product issues affecting thawing to the supplier for further corrective actions to be made at the source of the raw animal food production.

## Managers' Role in Training Employees

Certified food safety managers are the best equipped to train employees on safe food handling specific to the preparation of the menu of the foodservice business and the other SOPs associated with the Prerequisite Control Program. This is because the CFPM will monitor employee behaviors using the FSMS based on the food safety knowledge of the food handler employee and sometimes use re-training as a corrective action especially when employees do not follow the proper personal hygiene requirements in the FSMS. Managers may also influence employee safe food handling behaviors through their oversight/supervision using the FSMS as well. In a survey (Arendt and Sneed 2008) designed to measure employee motivators for following safe food handling practices (hand washing, wearing clean uniforms, cleaning and sanitizing properly, and measuring food temperatures properly), the roles of a supervisor that employees observe had the largest influence as motivators of proper safe food handling. For example, establishing policies, expecting accountability (via an FSMS), serving as a role model, providing training, providing awards, and providing resources were all reported as motivators.

## Role of Food Handlers

Fundamental to effective management of people who prepare and serve food is the need that they understand the requirements they must practice and procedures they are expected to follow to ensure safe food. Food handlers or non-managerial employees who prepare and serve food should be knowledgeable of the same basic knowledge that food safety managers must know, including causes of foodborne illnesses and how to prevent them using the controls defined in FSMS (e.g., managing their own health and personal hygiene). Many foodborne disease outbreaks are associated with employees who work and handle food while sick because of a lack of effective training (see above). For example, a food handler infected with Norovirus who works while sick and does not wash his/her hands properly and/or wear foodservice gloves while handling foods can infect a significant number of customers. Historically, sick food handlers who transmit their illness to others (employees and customers) through the food they prepare play a significant role in the root cause of almost half (46%) of all restaurant-associated foodborne disease outbreaks in the United States (Gould et al. 2013). Infected food handlers cause about 70% of reported Norovirus outbreaks from contaminated food (CDC 2014). Although FSMS will be designed to monitor and control this (which must be monitored daily to ensure proper employee-behavior), because some of the controls are

**Table of Contents**

Food Safety

**Importance of Food Safety** . . . . . . . . . . . . . . . . . . . . . . . . . . . . . 3
**Contamination Hazards** . . . . . . . . . . . . . . . . . . . . . . . . . . . . . . . 4
Types of Contamination . . . . . . . . . . . . . . . . . . . . . . . . . . . . . . 4
Cross-Contamination . . . . . . . . . . . . . . . . . . . . . . . . . . . . . . . . 8
**Personal Hygiene and Health** . . . . . . . . . . . . . . . . . . . . . . . . . 11
Washing and Sanitizing Hands . . . . . . . . . . . . . . . . . . . . . . . . 11
Wearing Food Service Gloves . . . . . . . . . . . . . . . . . . . . . . . . 14
Dealing with Illness and Injuries . . . . . . . . . . . . . . . . . . . . . . 16
**Cleaning and Sanitizing** . . . . . . . . . . . . . . . . . . . . . . . . . . . . . 18
Using Sanitizing Solution Correctly . . . . . . . . . . . . . . . . . . . . 18
Cleaning Up Body Fluids Safely . . . . . . . . . . . . . . . . . . . . . . . 20
Using Cloth Towels Safely . . . . . . . . . . . . . . . . . . . . . . . . . . . 24
Handling Trash Safely . . . . . . . . . . . . . . . . . . . . . . . . . . . . . . . 25
**Controlling Temperature and Time** . . . . . . . . . . . . . . . . . . . . 26
Dealing with the Temperature Danger Zone . . . . . . . . . . . . 26
Thermometers . . . . . . . . . . . . . . . . . . . . . . . . . . . . . . . . . . . 28
**Flow of Food** . . . . . . . . . . . . . . . . . . . . . . . . . . . . . . . . . . . . . 31
Primary Critical Control Points (CCPs) in the Flow of Food . . . . . . . . . . . . . . . . . . . . . . . . . . . . . . . . . . . . 31
Receiving/Storing Food Safely . . . . . . . . . . . . . . . . . . . . . . . . 33
Preparing Food Safely . . . . . . . . . . . . . . . . . . . . . . . . . . . . . . 35
Cooking/Assembling Food Safely . . . . . . . . . . . . . . . . . . . . . 37
Holding Food Safely . . . . . . . . . . . . . . . . . . . . . . . . . . . . . . . 39
Serving Food and Beverages Safely . . . . . . . . . . . . . . . . . . . 40
**Food Allergies and Intolerances** . . . . . . . . . . . . . . . . . . . . . . 42
Overview . . . . . . . . . . . . . . . . . . . . . . . . . . . . . . . . . . . . . . . . 42
Avoiding Cross-Contamination of Allergens . . . . . . . . . . . . . 45
Answering Questions about Allergens . . . . . . . . . . . . . . . . . 46

**Fig. 5.2** Recommended components of a food handler training course (e.g., foodservice employee training program) that non-managerial employees should be trained on to ensure a more effective FSMS in a foodservice business

measured as the behavior of the employee, training and re-training food handler-level employees is an absolute requirement to prevent foodborne illnesses and outbreaks.

An employee-level food safety training course, sometimes called a food handler course, should provide the general knowledge employees need to prepare and serve safe food, and will be aligned to the demonstration of knowledge that the CFPM must know (those included in Fig. 5.2). Food handler courses are offered by several groups with ANSI-accredited certification exams like those that provide managerial training discussed above. These courses provide a good foundation in basic food safety knowledge that then aligns to the knowledge of the controls used in FSMS. Each employee should demonstrate food safety education through curriculum-based evaluations if an ANSI-accredited course is not used and application of the important food safety knowledge during work (e.g., hand washing and proper glove use), overseen by the CFPM. Some foodservice businesses decide not to train food handler employees on food safety due to perceived labor cost, turnover rates, etc. However, when they neglect to train employees on the most critical controls to prevent foodborne illnesses (personal hygiene training alone would be better than no training), a FSMS is less effective, more time is spent on corrective actions/training, and there is likely more food waste and poor customer experience.

There is additional knowledge that employees need to have that is helpful when there are food safety emergencies. A good example of additional knowledge all employees should have in addition to the general food handler training described above concerns how the establishment requires employees to manage body fluid cleaning and sanitation (a new requirement in the most recent FDA Food Code; see Chap. 4). For example, when a customer or employee produces body fluids in a foodservice environment (e.g., vomits in the dining areas), pathogens in the body fluid can quickly contaminate food contact surfaces, leading to cross-contamination of foods—even if the employee who cleans up the spill wears foodservice gloves. Many of the pathogens found in human body fluids (e.g., Norovirus) are more resistant to common foodservice sanitizers, making it even more important that employees are prepared to clean up body fluids properly and safely to keep body fluids out of the food prep areas. Allergen training and risk is equally as important (King and Bedale 2018)

Some states (e.g., California; see California Restaurant Association (2012)) require food handler certification training (ANSI level) of all food employees working within a foodservice establishment. Inspectors require evidence of this training in the form of either a food handler card (like as a license to prepare food safely) or other certificates. In some cases, violation of this requirement can lead to fines up to $1000 or closure of the foodservice establishment until training requirements are met. The primary reason states like California establish these public health laws is because many foodservice businesses neglect to train their employees on food safety requirements for preventing foodborne illnesses known to be effective (e.g., not working during a Hepatitis A or Norovirus infection), and in order to better protect the public health, they establish the means to regulate and enforce training.

## Training Tools

Additional tools can be used to reinforce exposure of food handler-level employees to proper food safety procedures, including (1) job aids, e.g., a sign showing how and when to wash hands properly posted at each hand wash sink or at the restroom door as a reminder to wash hands before leaving the restroom; (2) educational materials e.g., newsletter or weekly compliance rate reports and review with the CFPM; or (3) contests and rewards for proper food safety behaviors observed during FSMS assessments, e.g., free dessert at break time. In one study of food handlers exposed to job aids in highly visible locations such as kitchen work areas and hand washing sinks, there was a significant increase (as compared to before job aids were posted) in hand washing and decrease in indirect cross-contamination events in 47 food handlers working in eight foodservice environments (Chapman et al. 2010).

It is well documented that knowledge is acquired by training, but education is only achieved if the knowledge can be demonstrated. We all remember being called up to the front of the high school classroom to complete a math problem on the chalk board (my worst nightmare after not studying my homework the night before). Of course food safety training is not nearly as terrifying to a high school student as math (at least to me), but the outcome of education or lack of it can lead to much more serious consequences for many other people when neglected. If owners/operators neglect their training on HACCP, and training for their food handler employees on food safety, and then only rely on CFPM training for their managers, they will experience less effective FSMS and "leave the door open" to more risk of causing foodborne illnesses and outbreaks due to the behaviors of the very employees that might prevent them (CDC 2007).

# References

Arendt S, Sneed J (2008) Employee motivators for following food safety practices: pivotal role of supervision. Food Prot Trends 28:704–711

Brown LG, Le B, Wong MR, Reimann D, Nicholas D, Faw B, Davis E, Selman CA (2014) Restaurant manager and worker food safety certification and knowledge. Foodborne Pathog Dis 11(11):835–843

Brown LG, Hoover ER, Selman CA, Coleman EW, Rogers HS (2017) Outbreak characteristics associated with identification of contributing factors to foodborne illness outbreaks. Epidemiol Infect 145(11):2254–2262

Bryan FL et al (1997) Surveillance of foodborne diseases III. Summary and presentation of data on vehicles and contributory factors: their value and limitations. J Food Prot 60:701–714

Burke AJ et al (2014) Do certified food manager knowledge gaps predict critical violations and inspection scores identified during local health department restaurant inspections? Food Prot Trends 34:101–110

California Restaurant Association (2012)California Food Handler Card, SB 602 requires workers to receive food safety training. Available via internet athttp://www.calrest.org/issues-policies/key-issues/food-safety/foodhandler/

Cates SC, Muth MK, Karns SA, Penne MA, Stone CN, Harrison JE, Radke VJ (2009) Certified kitchen managers: do they improve restaurant inspection outcomes? J Food Prot 72(2):384–391

CDC (2006)CDC endorses certification of food safety kitchen managers. Available via internet athttp://www.cdc.gov/nceh/ehs/EHSNet/resources/certification.htm

CDC (2007) Norovirus outbreak associated with ill food-service workers, Michigan, January–February 2006. MMWR 56(46):1212–1216. Also available via internet at http://www.cdc.gov/mmwr/preview/mmwrhtml/mm5646a2.htm

CDC (2016) Kitchen manager certification, an important way to improve restaurant food safety. CDC Infographic. Available via internet athttps://www.cdc.gov/nceh/ehs/docs/factsheets/ckm-infographic.pdf

CDC (2017a) CDC's national environmental assessment reporting system (NEARS) 2015 summary report. Available via internet athttps://www.cdc.gov/nceh/ehs/nears/docs/2015-nears-report.pdf

CDC (2017b) Infographic contributing factors: preventable causes of foodborne illness. Available via internet athttps://www.cdc.gov/nceh/ehs/publications/cf-infographic.html

Centers for Disease Control and Prevention (2014) Preventing norovirus outbreaks. Retrieved fromhttp://www.cdc.gov/vitalsigns/norovirus/index.html?s_cid=ostltsdyk_cs_500

Chapman B, Eversley T, Fillion K, MacLaurin T, Powell D (2010) Assessment of food safety practices of food service food handlers (risk assessment data): testing a communication intervention (evaluation of tools). J Food Prot 73:1101–1107

Food and Drug Administration (FDA) (2006) Managing food safety: a manual for voluntary use of HACCP principles for operators of foodservice and retail establishments. Available via internet athttps://www.fda.gov/downloads/food/guidanceregulation/haccp/ucm077957.pdf

Food and Drug Administration (FDA) (2010) Retail food protection: employee health and personal hygiene handbook. Available via internet at http://www.fda.gov/downloads/Food/FoodSafety/RetailFoodProtection/IndustryandRegulatoryAssistanceandTrainingResources/UCM194575.pdf

Food and Drug Administration (FDA) (2017) FDA Food Code. Recommendations of the United States Public Health Service, Food and Drug Administration, National Technical Information Service Publication number IFS17. Food and Drug Administration

Food and Drug Administration (FDA) (2018) FDA report on the occurrence of foodborne illness risk factors in fast food and full-service restaurants, 2013–2014

Gould LH, Rosenblum I, Nicholas D, Phan Q, Jones TF (2013) Contributing factors in restaurant-associated food-borne disease outbreaks, FoodNet sites, 2006 and 2007. J Food Prot 76(11):1824–1828

Hedberg C et al (2006) Systematic environmental evaluations to identify food safety differences between outbreak and nonoutbreak restaurants. J Food Prot 69:2697–2702

King H, Bedale W (2018) Managing food allergens in retail quick-service restaurants. In: Food allergens: best practices for assessing, managing and communicating risk. Springer, Cham

# Chapter 6
# Facilities That Enable Food Safety Management Systems Execution

Design specifications of primary food storage and preparation areas are key to enabling Active Managerial Control of food safety hazards using Food Safety Management Systems (FSMS) in a foodservice establishment. These specifications are necessary to ensure that (1) each process step can be designated to the location and (2) their associated hazards can be controlled by barriers (e.g., separating areas where raw proteins and ready-to-eat (RTE) foods are prepared). Standard operating procedures (SOPs) such as hand washing, glove use, and cleaning and sanitation tools necessary to prevent cross-contamination and the flow of foods from raw to RTE food preparation to service are the goal of facility design to support Process HACCP–based FSMS that keep food safe. This facility design facilitates the separation and intuitive movement of employees and food preparation activities during the processes without unnecessary crossover of task and space (see Fig. 6.1). Facility design should include exclusion of pest (flies/roaches/rodents) as well as non-employees/customers who should be restricted from food preparation or storage areas, and provide for proper food storage space located at the point of receiving (approved source of) foods from outside the facility.

## Facility Design for Reduction of Risk

The FDA (FDA 2000) provides general guidelines on the facility design that a foodservice establishment should have to ensure food safety, including:

- Materials for construction and repair

  - Cleanability (e.g., floors, walls, and ceilings)
  - Functionality (e.g., protective shielding of light bulbs, heating, ventilating, air conditioning vents, insect control devices and proofing, restrooms, closures)

© Springer Nature Switzerland AG 2020
H. King, *Food Safety Management Systems*, Food Microbiology and Food
Safety, https://doi.org/10.1007/978-3-030-44735-9_6

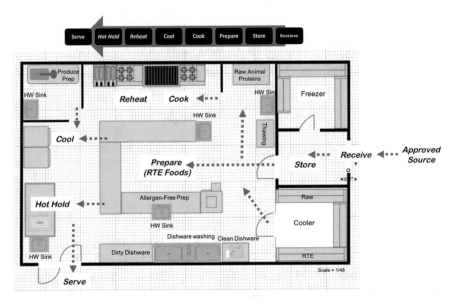

**Fig. 6.1** Safe flow of food through each expected process in a foodservice facility, designed to enable Process HACCP controls of a same-day prep for service or complex process menu (note that hand wash sinks would have splash guards on all sides—not shown)

- Numbers and capacities

  - Availability of hand washing cleansers
  - Hand drying provisions
  - Hand washing signage
  - Lighting intensity
  - Dressing areas and lockers/employee food storage

- Location and placement

  - Hand washing sinks
  - Restroom convenience and accessibility
  - Segregation of damaged, spoiled, or recalled products
  - Placement of waste receptacles

- Maintenance and operation

  - Cleaning frequency
  - Drying mops
  - Closing restroom doors
  - Controlling pests
  - Storing maintenance tools
  - Prohibiting animals

However, the local regulatory authority will always determine the final design requirements (which will be required to be specific to the menu prepared and served) and approval of any foodservice facility used for business operations. These requirements are normally based on defined specifications/building code rules set by the state and/or local authorities that include the sanitary design, facility layout, operational and product flow, menu use, construction materials, equipment, plumbing, and facility surroundings. For example, the state of Georgia has a comprehensive plan review process (which is representative of most other state rules aligned to the FDA Food Code), including step-by-step procedures for foodservice operators to complete in order to design new or renovate existing facilities for operating a foodservice establishment (see *The Food Service Establishment Manual for Design, Installation and Construction*, Georgia Department of Public Health, Environmental Health, Food Service 2013). The 437-page manual provides detailed specifications for each of the areas described above, including additional equipment specifications and templates that assist in the application process for new or renovated facility construction. The manual provides example foodservice facility types and scenarios and detailed information, including explanations of the necessity for each requirement to ensure food safety with relevant case studies. This manual is also a good resource for learning more about design of foodservice faculties based on FDA-aligned food code requirements, including use of Process HACCP–based FSMS with minor revisions to meet any states rules. It demonstrates regulatory authority expectations during the plan review approval process and operations (i.e., health department inspections will be performed to verify facility design specifications). Some local and state regulatory authorities also provide a quick reference summary of the minimum requirements to guide foodservice facility design planning that can also be helpful in understanding the additional expectations of local regulatory building code rules, including requirements, for example, a mop sink, hot water storage capacity, certified equipment for foodservice, plumbing for service of carbonated beverages, door sweeps, etc.

An example list of minimum requirements for a typical foodservice and beverage establishment that a state may require for a pre-opening plan review could include a requirement for submission of (Source: Maricopa County Environmental Services Department 2017):

1. The intended menu for plan review. A finalized menu is required for the regional/program office file.
2. A plan to protect open foods.
3. A plan for a mop sink/service sink within 300 ft of the kitchen.
4. A plan for hand sinks to be centrally located, visible, and directly accessible within 25 ft of all food and beverage work stations:

    (a) Waterproof metal splash guards, at least as high and as wide as the hand sink, shall be installed between the hand sink and food/beverage/utensil-related areas within 24 in. of the hand sink (this is important when hand sinks are next to raw animal protein preparation).
    (b) Food caterers shall have approved accessible hand washing facilities that shall be located and maintained open at all times.

(c) Where suitable facilities are not immediately accessible, an approved self-contained hand washing station will be provided.

5. Evidence of a National Sanitation Foundation (NSF)-approved 3/4-compartment sink to wash dishware. The 3/4-compartment sink drain line shall be indirectly plumbed with a minimum 1-in. air gap where permissible.

6. Evidence that all sinks including mop/service sinks shall have mixing faucets or valve.

7. Evidence that if an NSF-approved commercial dish washer is installed, an approved three-compartment sink is also provided.

8. Evidence that a separate NSF-approved stainless steel food prep/rinse sink indirectly plumbed with an approved minimum 1-in. air gap off the drain line for both defrosting food or rinsing produce.

9. Evidence for a minimum 50-gallon water heater with a 100% recovery rate.

10. Evidence for a minimum 75-gallon water heater required with a 100% recovery rate for an establishment with a mechanical dish washer.

11. Evidence for a "tank-less" water heater may be approved, but requires prior review.

12. Evidence for hot water or tempered water (100–110 °F) tap to all sinks.

13. Evidence for an audible or visual alarm for a mechanical sanitizer dish washer.

14. Evidence that employees will be provided with toilet facilities/restrooms in all occupancies; employee toilet facilities shall be either separate or combined employee/public toilet facilities.

15. Evidence that restrooms will be within 300 ft in covered malls and 500 ft in occupancies other than covered malls.

16. If the establishment serves carbonated beverages, evidence that an approved testable reduced pressure backflow preventer is properly installed between the water supply line and the carbonator.

17. Evidence that all equipment and fixtures connected to a water supply shall be equipped with an approved backflow prevention device or air gap to prevent backflow. Ice machines are typically exempt.

18. Evidence that garbage disposals are not installed on food preparation sinks, 3/4-compartment sinks, but may be allowed on separately plumbed pre-rinse/scrapping sinks or dedicated vegetable waste grinders.

19. Evidence that all equipment is NSF/ANSI-certified, approved equivalent commercial equipment, or equipment deemed acceptable by the department; additional equipment requirements will be determined by proposed menu type.

20. Evidence that all finishes, lighting, plumbing, and ventilation meet current building code department requirements.

21. Evidence that approved millwork shall be smooth and sealed and all food preparation surfaces are stainless steel.

22. Evidence that exterior and toilet room doors shall be installed and self-closing and door sweeps, thresholds, and weather stripping are installed at all exterior doors.

23. Evidence that all continuous openings/doors to the exterior shall be properly protected with approved air curtains or screens to prevent insect intrusion.

Another source of valuable and regulator supported criteria for foodservice facility design is *The Food Establishment Plan Review Manual,* produced by the Conference for Food Protection (CFP 2016). This resource is a training program available to foodservice operators who are seeking to build or remodel foodservice facilities to enable safe food production, and to prepare for the plan review process. This manual was written by foodservice and regulatory experts on the CFP Plan Review Committee (based on the FDA Food Code) to assist regulatory authorities, architects, food safety consultants, and industry facility design departments. Below is a summary of the items that should be defined in preparation for the design of a foodservice facility, including outside the primary food storage and food prep areas (for details, review CFP 2016):

- Proposed menu, seating capacity, and projected daily sales volume for the foodservice facility.
- Provisions for adequate rapid cooling, including ice baths and refrigeration, and for hot and cold holding of Time/Temperature Control for Safety (TCS) food.
- Location of all food equipment. Each piece of equipment must be clearly labeled, marked, or identified. Provide equipment schedule that identifies the make and model numbers and listing of equipment that is certified or classified for sanitation by an ANSI-accredited (like NSF) certification program (when applicable). Elevation drawings may be requested by the regulatory authority.
- Location of all required sinks: hand washing sinks, ware washing sinks, utility sink, and food preparation sinks (if required).
- Auxiliary areas such as storage rooms, garbage rooms, toilets, basements, and/or cellars used for storage or food preparation.
- Entrances, exits, loading/unloading areas, and delivery docks.
- Complete finish schedules for each room including floors, walls, ceilings, and covered juncture bases.
- Plumbing schedule including location of floor drains, floor sinks, water supply lines, overhead waste-water lines, hot water-generating equipment with capacity and recovery rate, backflow prevention, and wastewater line connections.
- Location of lighting fixtures.
- Source of water and method of sewage disposal.
- A color-coded flow chart may be requested by the regulatory authority demonstrating flow patterns for:
  - Food (receiving, storage, preparation, and service)
  - Utensils (clean, soiled, cleaning, and storage)
  - Refuse (service area, holding, storage, and disposal)
- Storage of employee personal items.
- Ventilation.

## Menu Impact on Facility Design

The ingredients used to prepare food products and the menu served by a foodservice establishment will influence what space, flow of foods, and equipment are required to facilitate FSMS to control all the potential hazards at each place and time that they are likely to occur. For example, consider the following scenario: a foodservice establishment prepares dozens of food products using multiple ingredients to provide a menu of baked chicken, hamburgers, leafy green salads with boiled eggs, potato salad, deli meats, fried chicken, and cold chicken salad sandwiches. All animal proteins are received raw; many of the final products are held cold or hot during service; some final ingredients/products require cooling after cooking or reheating after cold storage. Further, many of the raw animal foods are received frozen, some are received cooked, and all of the produce menu items are prepared fresh daily from bulk raw produce ingredients. Each of the food ingredients themselves (acquired from approved sources) may have unique hazards associated with each ingredient type (e.g., *Salmonella* from eggs, *E. coli* O157 from raw ground beef, *Listeria* from leafy greens, *C. perfringens* from raw chicken). These hazards may increase risk (FDA 2009) of cross-contamination to surfaces (e.g., refrigerator handles, cook equipment buttons, hand washing sink handles) as the foods are taken through the flow of food processes (see Fig. 6.1). In addition, other potential hazards arise from not cooking the food to the proper temperature, employees not washing hands, or not cooling down the cooked chicken properly by time and temperature, for example.

## Facility Design for the Process HACCP Plan and Prerequisite Control Program

### *Facility Design for Receiving*

It is important to ensure approved sources of food ingredients and products are used which includes a critical control in FSMS for rejecting (and documenting) any unapproved sources of foods or approved sources that do not meet food distribution requirements (e.g., cold food maintained and received at or below 41 °F) and prohibiting them from being delivered to the establishment. This is an important daily monitoring process in a FSMS (see Chap. 4) that can reduce food safety hazards before a food is received into the facility, as would be the case if fresh RTE foods were delivered above 41 °F, or if the containers had evidence of pest infestation or exposure.

Once food has been determined to be acceptable for receiving, the facility should be designed to accommodate the volume of food ingredients and products being delivered on a daily basis. Although the receiving area likely cannot accommodate separation of food types during the receiving process (e.g., raw and ready-to-eat foods separation), the receiving area should provide sufficient space to bring the foods into the facility from the delivery vehicles and protect them from the elements. Likewise, food items should not be received into the facility unless there is proper storage capacity after (and close to) receiving to ensure all raw ingredients can be stored separate from ready-to-eat foods in dry, refrigerated, and frozen storage. Foods can be contaminated during the receiving process before proper storage due to temperature abuse, dust, spillage of chemicals in use in a mop sink, etc.

Another important consideration is if and when "key-drop" deliveries of food are allowed. Key-drop deliveries occur when a third-party distributor is allowed access to the facility during non-operating hours (e.g., after closing) to make deliveries to the foodservice establishment. The risk associated with these types of deliveries is that often the items are "received" after business hours without verification by employees of the foodservice establishment that the products meet food distribution requirements, such as critical controls of approved product vendors and cold temperature compliance. Of course, this function can be retrospectively performed by employees during the next operating hours of the business (e.g., in the morning prior to opening the business), but the third party should be required to check and record temperatures of received food product and ensure inspection for damage/pest infestation (to be verified/documented each delivery). If key-drop deliveries are an anticipated need of the foodservice business, it may be beneficial to also design a dual exterior and interior lock at entry into the delivery area to allow access to storage for key-drop deliveries by third parties while excluding access to the rest of the foodservice facility (e.g., see outer and inner door configurations at the receive/store areas in Fig. 6.1).

## Facility Design for Storing

Proper food storage, which includes SOPs for proper food rotation (i.e., date of expiration tracking, first-in-first-out (FIFO), and date/time of all food deliveries) while food is in storage, is an important prerequisite to preparing and serving safe food. First, the facility design should ensure dry, cold, and frozen storage space capacity to meet both food rotation requirements and the volume of ingredients/products needed to meet product sales volume of the foodservice establishment. Ensuring that foods "flow" through storage according to their shelf-life time is also important to reduce food waste and resulting costs. Many food ingredients are perishable; thus, they must be used within short time periods after receiving. Storage space should be designated to place "use first" foods close to the front of the space with easy access for employees, thereby ensuring FIFO. For example, if a plat of fresh strawberries is delivered bi-weekly to the facility but strawberries in current storage have not been completely depleted, storing the new delivery of strawberries behind the previous plat of strawberries (e.g., via storage rack system, Fig. 6.2) would encourage employees to use the more easily accessible—older—strawberries first, until depleted. Likewise, storing food should include easy-to-read date of expiration/prep labels on food containers (especially prepared foods), which

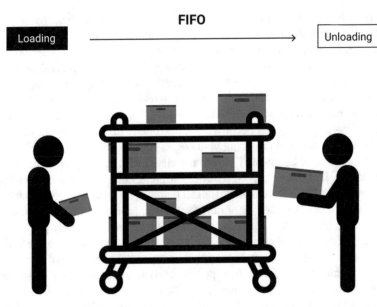

**Fig. 6.2** Example of racking in storage space to enable loading products with longest shelf life (from receiving) and pulling oldest products for food prep/sales using FIFO. (Source: Josh King)

facilitates FIFO and prevents accidental use of expired foods, which is a food safety hazard. Using color-coded date labels that identify the time and day of the week can help; however, when the storage space is small and crowded with food containers, it becomes difficult to ensure FIFO and food safety with only labels as they cannot be seen easily to alert employees to each item's food rotation requirements.

Facilities should be designed to accommodate complete separation of raw animal and ready-to-eat (RTE) foods during storage. For each foodservice establishment, this will depend on (1) the menu and requirements for extensive food preparation (washing, cutting, mixing, cooking, baking, frying, grilling, cooling, and holding) and (2) the capacity of storage needs (volume of sales per day and number of deliveries of food items per week). It is best to designate space for raw animal foods separate from RTE foods, especially when they are kept in the same storage cooler (e.g., designated sides of a walk-in cooler; see Fig. 6.1). Although the FDA Food Code requires all foods to be covered during storage to protect the food from contamination, not all covers/lids completely prevent spillage or dripping from the container (a clear reason to ensure that refrigeration units don't drip condensation on food containers). Because some foods are dispensed from food containers when in storage—for example, when an employee opens a food container to scoop a portion of the product into a cup for single service—there is also a higher risk of cross-contamination of RTE foods that are opened and closed frequently. Obviously, raw animal foods should always be stored below RTE foods in coolers and freezers (see Fig. 6.3); however, there is much lower risk of cross-contamination when they can be stored separately from RTE foods in a separate refrigerator as employees con-

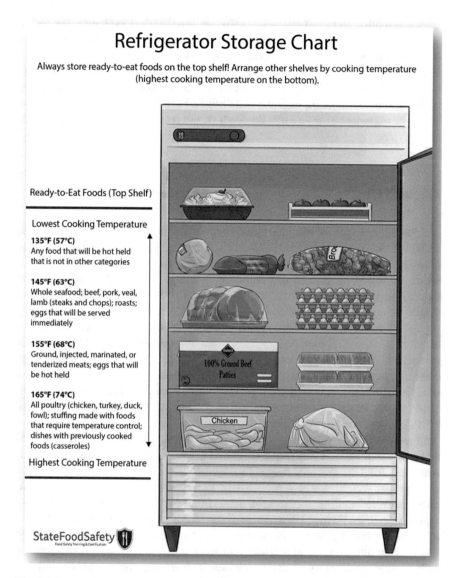

**Refrigerator Storage Chart**

Always store ready-to-eat foods on the top shelf! Arrange other shelves by cooking temperature (highest cooking temperature on the bottom).

Ready-to-Eat Foods (Top Shelf)

Lowest Cooking Temperature

**135°F (57°C)**
Any food that will be hot held that is not in other categories

**145°F (63°C)**
Whole seafood; beef, pork, veal, lamb (steaks and chops); roasts; eggs that will be served immediately

**155°F (68°C)**
Ground, injected, marinated, or tenderized meats; eggs that will be hot held

**165°F (74°C)**
All poultry (chicken, turkey, duck, fowl); stuffing made with foods that require temperature control; dishes with previously cooked foods (casseroles)

Highest Cooking Temperature

100% Ground Beef Patties

Chicken

StateFoodSafety
Food Safety Training & Certification

**Fig. 6.3** Proper storage placement of refrigerated food to prevent cross-contamination. (Source: State Food Safety)

tinually place and remove both raw and RTE foods from storage. This designated space for separation use (habit) provides repetition to the employee training and awareness of the risk of raw animal/fish food cross-contamination to establish consistency of use throughout the facility (see below).

A growing need in many foodservice businesses is that for safe storage of RTE ingredients/products for donation to food banks (segregated to exclude the products from preparation and sale until picked up or delivered to the food bank). The majority of foodservice facilities are not designed for storage of food items to be donated (nor are there any specific requirements for such storage by most regulatory authorities), but you can segregate/designate storage space for these items (through signage, as part of their FIFO plan). Providing a specific location for the safe storage of foods to be donated, for example in a designated refrigerator or freezer location, would be important to ensure that the foods do not reenter the flow of food for prep and service.

## Facility Design for Preparing, Cooking, and Reheating

If a foodservice facility will prepare raw animal and/or fish foods, food prep areas should be designated to facilitate separation of raw animal/fish prep from RTE food prep during operations (see designated areas, Fig. 6.1). Oftentimes, raw animal foods are stored frozen and therefore also require a designated space for proper thawing to reduce risk of cross-contamination. Cross-contamination can be prevented using the same space and equipment separated by time (as is allowed by regulatory authorities). To achieve this, environmental surfaces and/or equipment are used first for RTE food prep, properly cleaned and sanitized, and then used for raw animal food prep. However, time constraints can necessitate dual use (e.g., catering prep, unexpected large orders, change in menu item recipe, special orders of RTE foods, etc.). A designated separation space/area for raw animal food preparation mitigates risk of overlapping use of food preparation surfaces, which also provides repetition (through consistency of use of separation in storage and food prep areas) to employees' awareness of risks associated with raw animal/fish food cross-contamination. It also facilitates consistent use of the proper flow of raw animal foods from the raw food prep area to the cooked RTE area.

Improper facility design can contribute to inadequate protection of food from dirty dishware handling, for example, due to improper placement of produce preparation sinks in the facility. For example, if produce is rinsed and cut in a sink next to a 3/4-compartment dish washing sink where dirty dishware like pans used to store raw chicken may be sprayed off before washing, it can lead to cross-contamination of RTE produce. Produce (fruits, vegetables, leafy greens) that are not cooked further are always considered RTE; therefore, if a foodservice business prepares a large proportion of its menu using raw bulk produce (e.g., whole heads of romaine lettuce or cabbage), it would be best to provide a designated area/room for produce preparation. Bulk produce should be washed and handled separately from all raw animal foods and other ingredients not considered RTE (e.g., even flour which is considered a raw agriculture ingredient not a RTE food), if daily menu needs require produce prep.

Allergen-containing ingredients used in food preparation are equally important to keep separate from foods made for customers to be allergen-free (e.g., peanut-free product). Specified food preparation areas should be defined and used to help manage food allergen cross-contact—especially if the business communicates to the customer its ability to prepare foods free of specific allergens. Recipes can exclude the top eight allergens so that none of these allergens should be present in the restaurant prep kitchen. Alternatively, the allergen (e.g., almonds) can be packaged by a supplier in a separate package (e.g., like a condiment) and then given to the customer to add to their product after purchase to avoid need for separation. Moreover, color-coded tools (e.g., purple for allergen-free zones only) can be provided to be used exclusively in designated prep areas, thus reducing the cross-contact of allergens to other foods (King and Bedale 2018).

Because many food items are cooked from raw ingredients and/or raw animal/fish foods in the cooking area, it is important to establish a "raw-to-cooked" flow for the food process in this area. This is especially true when the raw food item is removed directly from a raw food container (e.g., marinated raw chicken) and placed directly on the cook surface (e.g., grill). Many of the raw foods and/or containers likely will come into contact with cook equipment surfaces just before the items are placed to be cooked (e.g., if an employee places the raw food container on the non-cooking surface area to enable loading the cook surface). When raw foods contain breading or other coatings that can easily cross-contaminate non-cooking area surfaces, it is important to establish the proper flow for (1) loading raw food to be cooked and then (2) unloading cooked RTE foods in these areas to a separate place for holding RTE foods. When foods need to be reheated for food safety, it is best to use cook equipment specific for reheating rather than using the same equipment to also cook raw animal/fish foods (if feasible) to reduce crossover of different tasks in this area. Many foods that can be reheated by microwave or a conventional oven make this separation more feasible.

## Facility Design for Cooling, Hot Holding, and Serving

The best design for the food cooling and hot holding processes within a foodservice facility involves the use of designated equipment for each process, located in the areas where RTE foods are stored and/or served. For example, when foods need to be cooled down after cooking according to requirements (e.g., below 41 °F within 6 h total), a blast chiller equipment that records time and temperature is the best solution; the chiller can also serve as the cold holding location before service. If a blast chiller is not available, an upright refrigerator can be used along with a temperature monitoring system to track temperature changes during food cooling. When foods must be held hot prior to service (non-buffet or self-service), it is best to use hot holding equipment located in close proximity to the service area where employees can easily dispense the foods away from food prep—especially the raw animal preparation areas. It is important to always verify that foods being cooled down or held hot are at the correct temperature (i.e., not in the temperature "danger zone" between 41 and 135 °F).

There are numerous types of foodservice in numerous different business concepts, for example, quick-serve, fast-casual, full-service dining, etc. These also include cook-to-order (or immediate service), self-serve (buffet or salad bar), prepackaging for delivery or pick-up, large-volume service (catering or institutional), and now food preparation for third-party (non-employee) pick-up and delivery. The need for service area design in a foodservice facility is normally based on these types of service needs. Because the food served will always be RTE foods (raw fish products like sushi are still considered RTE), it is best to design the food service areas (holding, staging, plating foods, etc.) at the area in the facility in closest proximity to the point of customer pick-up of consumption (buffet, order line, or dining area). It is important to provide enough space for these foods in service to prevent not only co-storage of dirty dishware (e.g., being returned from full-service tables), cleaning tools, or other non-food items but also reduce crossover of tasks including cleaning and sanitation (e.g., not near the mop sink). This area should be the cleanest area of the foodservice facility at all times (next to the dining area) and enable the free movement of employees into and out of the kitchen area. Obviously, the same can be said for the dining areas where customers consume the foods after service. Additional considerations of the safe food service process and hospitality are discussed in more detail in Chap. 4.

## Enabling Hand Washing to Prevent Hazards During Food Preparation Processes

Access to and provision of adequate hand washing equipment/sinks is another important facility design-related need. Hand washing sinks should be logically integrated into the flow of the food prep and service design. Hand washing is the most critical means to reduce pathogens on hands (and thus cross-contamination to ready-to-eat foods). Thus, each hand washing station should form a "hurdle" between raw animal foods and RTE foods including the processes of cooking; meaning that employees can easily wash hands between food prep task in the two, physically separated areas. It is equally important to have a hand wash sink near the dishware washing area, so employees can easily wash hands between the task of handling dirty dishes (loading) and handling clean dishware (unloading/storing).

Because one of the Prerequisite Control Points (PCP) to prevent foodborne diseases is associated with each team member's health and personal hygiene (e.g., employee who has undiagnosed Norovirus infection), it is also important to ensure a hand washing sink is located at the primary entry point for employees as they enter the food preparation area—for example, the "back-of-the house" or the kitchen—so that employees can wash their hands after going into and out of the kitchen; first in the restroom itself and again before entering the kitchen (a double hand wash requirement whenever they leave the kitchen to go to the restroom). A best practice SOP is to require all employees to perform a hand wash every time they return to the kitchen no matter what the reason is for leaving the kitchen (e.g., to empty trash, clean dining area tables, etc., and even the first time they clock into work). Likewise, this hand wash sink can be used to ensure that employees wash their hands before they leave the kitchen to serve customers.

Location of hand washing sinks away from food preparation areas (including ice and beverage handling/prep) is also an important design consideration in the safe flow of food to ensure that hand washing activities themselves do not cross-contaminate food. If a hand washing facility must be located near any food prep, a stainless steel barrier (with height defined by risk of splash from hand washing—generally high and wide enough to prevent splash over into foods) should be installed.

## Design for Cleaning and Sanitation

The FDA Food Code and most state regulatory foodservice facility plan review documents include requirements for the cleanability surface characteristics of floors (and grout between tiles), walls, ceilings, food prep tables, food contact surfaces, sinks, etc. They also include restrictions on some surface types such as floor carpeting. Equipment (ANSI-certified) placement in the design is also critical to ensure cleanability, especially where the surface meets the walls or other equipment/tables. Environmental surfaces in a foodservice establishment can harbor food debris, which can lead to the development of biofilms of biological hazards (e.g., *Listeria monocytogenes* can produce a biofilm on stainless steel surfaces and then grow to larger numbers of bacteria on the surface). Thus, all surfaces must be accessible for cleaning and sanitizing to remove all biological hazards even if not a food contact surface. If these surfaces are neglected, they can lead to cross-contamination as the biofilms accumulate and are released on food contact surfaces or touched by employees that can then transmit the biofilms to other food preparation surfaces or food. This is especially important in dishware storage and washing area surfaces, including the racks and shelving used to store dirty dishware. Also, dirty dishware cleaning and storage areas must be adequately separated from clean dishware and adequate space must be available for drying dishware. Otherwise, a single cross-contamination event due to stacking wet dishware—for example, a dirty container that held raw chicken comes in contact with a clean container that will later be used to store prepared RTE chicken salad could easily cause a large outbreak of food-borne illness. A best practice is to design the clean dishware storage area with racks and slots that will accommodate each and every individual dishware item to drip dry without stacking.

Another important facility design requirement to ensure proper cleaning, sanitation, and safety of the foods prepared in the facility concerns the location of the mop/utility sinks. Because these sinks are used for preparing chemicals for cleaning and/or storing cleaning tools and chemicals, they should be located away from food storage, prep, and customer service areas (preferably outside the kitchen area; for example, see Fig. 6.1). If it is necessary that the sinks be located inside the kitchen area, they should be "guarded" by splash guards that prevent splashing of any solution/chemicals onto other surfaces, especially dishware and food prep areas.

## Preventing Pest Infestations

Facility proofing—elimination of gaps and entryways used by pests to enter a foodservice facility—is essential for any business storing and preparing foods. Proofing a structure involves identifying any portal that a pest can potentially use to enter a location and limiting access via that portal. This can be as tiny as gaps in the molding between walls and floors or as large as holes in walls behind equipment. Proofing can be achieved through measures such as using an inexpensive caulk to seal gaps between wood molding at floors to prevent insect entry or wire mesh to close gaps in walls where pipes and electrical conduits occur to prevent entry of rodents. This can be especially critical for foodservice facilities in malls or other buildings that certain share spaces with other businesses. For example, shared ceiling spaces between foodservice facilities (i.e., no walls that separate the facilities between the ceiling and roof) can allow pest movement and infestation.

Besides proofing the facility, consideration should also be given to how the landscaping is configured around the facility. Any vegetation touching the roof or walls of the structure should be trimmed back. Ideally, organic mulches should be replaced with rocks or stone. Air curtains (equipment that blows air away from window and door openings to reduce fly entry) and door sweeps on the bottom of any door open to the outside of the kitchen areas can be very helpful in preventing pest entry into the facility. Keeping all trash and litter in sealed containers (and stored away from food prep areas when not in use) will also make the facility much less attractive to rodents and insect pests. The key to the foodservice facility design for food safety is to prevent any pest infestation from occurring. Many pests carry biological hazards such as *Salmonella*; their movement within the facility can leave these pathogens (and their oil + urine) on food contact surfaces that may look clean but are not and then when used to prepare foods, will lead to a foodborne illness.

## Signage and Job Aids in Design

Finally, a kitchen that is designed to enable the Process HACCP FSMS to accommodate the safe flow of foods (in which the foods "move" in a natural flow of processes through the establishment ending with the holding and serving of RTE foods to customers; see Fig. 6.1) ensures that employees know where each process should occur. This establishes proper procedures for control of hazards during each process that, once employees have been trained, will become ingrained through repetition as habit. This design also reduces the risk of cross-contamination by employees (once they consistently wash their hands properly at the appropriate times) and establishes a routine of safe food storage and handling in the facility. An important addition to this design is signage and job aid placement to help employees navigate proper use and behaviors while working in the facility. The most common signage used in foodservice facilities (required in most states) is the "how to and when to wash hands" or "employees must wash hands" signage found at hand wash sinks or on the inside of doors in restroom facilities.

Signage placement in a facility can serve a similar function as road signage (like "CAUTION" or "YIELD" signs) that helps drivers drive safely on roads (for those who pay attention to signs, of course). For example:

- A "no bare-hand contact" sign can function like a **STOP** sign on roads, advising caution to the employee that they must not handle foods with bare hands in this area.
- A "raw chicken/animal proteins only" sign can function like a **NO PASSING** sign to warn employees not to "pass" this space in the safe flow of foods if handling raw animal proteins.
- A "no sick employees" sign can serve as a **NO ENTRY** sign, warning employees not to work if sick.
- A "three steps in dish washing" sign can serve as a **ONE-WAY** sign, where dish washing flow should only go one way (from dirty to clean to storage).

I often wonder if we should use "road signs" in foodservice to better caution employees and guide the flow of food in the correct direction. The International Association for Food Protection (IAFP) provides a set of well-designed icons/signage that can be used as cautionary/reminder signage throughout a facility as job aids for the proper location of food preparation (available for no cost; see International Association for Food Protection 2018), as demonstrated in Fig. 6.4. These icons can be published on any paper or plastic material and posted in respective designated areas to reinforce the proper food safety requirements of that area; they can also be used in all training materials and recipes to enable link to the requirements to the areas in the facility.

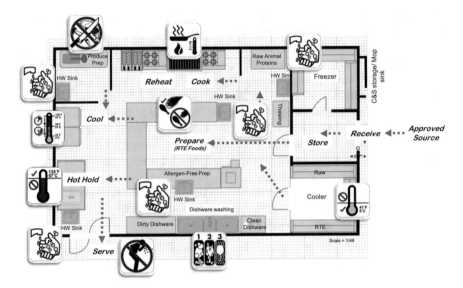

**Fig. 6.4** Example of placement of signage to note food safety SOP or area use requirements in a facility designed for Process HACCP. (Food Safety Icons Source: International Association for Food Protection, copyright)

# References

Conference for Food Protection (2016) Food establishment plan review manual. Available via internet athttp://www.foodprotect.org/media/guide/2016-plan-review-manual.pdf

Food and Drug Association (FDA) (2000) Food establishment plan review guide. Available via internet at http://www.fda.gov/Food/FoodSafety/RetailFoodProtection/ComplianceEnforcement/ucm101639.htm

Food and Drug Association (FDA) (2009) FDA report on the occurrence of foodborne illness risk factors in selected institutional foodservice, restaurant, and retail food store facility types (2009). Available via internet at http://www.fda.gov/Food/FoodSafety/RetailFoodProtection/FoodborneIllnessandRiskFactorReduction/RetailFoodRiskFactorStudies/ucm224321.htm

Georgia Department of Public Health, Environmental Health, Food Service (2013) Design, installation and construction manual. Available via internet athttps://dph.georgia.gov/food-service-design-installation-and-construction-manual

International Association for Food Protection (2018) Food safety icons. Available via internet athttps://www.foodprotection.org/resources/food-safety-icons/

King CH, Bedale W (2018) Managing food allergens in retail quick-service restaurants. In: Food allergens: best practices for assessing, managing and communicating risk. Springer, Cham

Maricopa County Environmental Services Department (2017) Plan review, minimum requirements for establishments. Available via internet athttps://www.maricopa.gov/DocumentCenter/View/5797/Minimum-Requirements-PDF

# Chapter 7
# Digital Technology to Enable Food Safety Management Systems

*The future of food safety is DIGITAL*
Steven A. Lyon, Ph.D., Chick-fil-A

## Why Digital

Digital technology, combined with a commitment to transparent self-assessments/corrective actions, is key to the most effective Food Safety Management Systems (FSMS) that will provide assurance of Active Managerial Control in a restaurant. We have all experienced the impact of digital technology (often in the form of mobile apps on a smartphone) through the use of these technologies to manage our daily work. Restaurant employees may not normally associate digital technology applications like mobile apps on a smartphone with food safety management in a restaurant, though they frequently use mobile apps for many other similar purposes at work and home. Most restaurant businesses use paper checklists to perform daily self-assessments of quality, inventory, work schedules, etc., and only use checklist for limited monitoring of food safety controls such as for checking the temperature of foods. However, the need for digital technology to enable the necessary comprehensive management of food safety controls in a restaurant using FSMS is significant because of these:

- FSMS SOPs must be documented including a means to capture the data from the monitoring functions easily.
- Multiple different specifications are required to monitor each control properly (e.g., what temperature should pork be cooked to), and there is a requirement to provide the required corrective action specific to that control at the time of the assessment.
- There is a need for scheduling of assessments (via calendar) and having access and using the work schedule to manage employee labor that may be impacted when employees are sent home due to illness; tracking employees on a sick log

© Springer Nature Switzerland AG 2020
H. King, *Food Safety Management Systems*, Food Microbiology and Food Safety, https://doi.org/10.1007/978-3-030-44735-9_7

is more effective to ensure sick employees excluded from work do not return to work until well.

- A need to track training needs including re-training of employees as a corrective action (e.g., observed not washing hands properly multiple times).
- There are communication needs (text and email alerts) of the results that may need immediate corrective actions performed by owners/operators or notification (i.e., exception reporting) when a corrective action is not performed, or a root-cause assessment is needed (leaving the hazard in place if not resolved quickly).
- A need to document an issue with photographs and text entry.
- A need to access the latest information/knowledge (Internet source of most current information) and ensure SOP updates are current.
- A need for the visibility of information to other managers between shifts.
- Digital technology can track and report false data entry (pencil whiping date, see below).
- A need for the means to quickly analyze information to make decisions and improvements in real time.
- A means to track trends in food safety assessment data over time to determine process execution issues (e.g., a cooling SOP cannot be executed properly using current equipment and procedures).

In contrast, when each of these tasks is attempted using a paper-based manual/SOP and paper-based checklist, the only realized value is documentation (if the checked item evaluations are performed properly and written down). Paper-based checklists have very limited visibility and searchability by all of the required stakeholders in the business. Further, they are limited to the location and require significant time for users to read through all of the data to find exceptions. Oftentimes a paper checklist based on an important process, recipe, or procedure is kept on paper in a three-ring notebook/manual stored in the restaurant office. They are often not current and not easily accessible. Paper-based checklists are not "active" in the sense that they can auto-analyze/report information or prescribe corrective actions when a process is not in compliance with the most current standard. They also cannot report and/or auto-alert management that an important control of a food safety hazard is not working.

An important aspect of FSMS is in the process of actively managing the food safety risk by evaluating whether proper controls are in place and immediately making corrective actions to prevent food safety issues. Oftentimes, a paper-based checklist also includes multiple quality checks or is in the form of multiple paper lists (e.g., one hanging in the restroom to remind employees to check-off when the restroom was cleaned last), which, together, dilutes the commitment to complete each properly. For example, when there are so many individual items on one or more checklists and/or little time available to check items, it is a tendency for untrained and/or hurried/stressed employees to "pencil-whip" (i.e., write the expected information) the information by not measuring the item and/or not entering actual information—even when there is a requirement to enter the name or initials of the person performing the evaluation.

Likewise, because not every employee (especially non-managerial staff) has the proper food safety training, a paper-based checklist cannot serve as a "tutor," describing the what, when, where, and how to ensure proper monitoring of controls. By contrast, a digital solution using a mobile app can inform the user with a mobile computer "at their fingertips" with this information:

1. What food safety controls need to be checked and when (e.g., temperature of chicken after cooking)
2. How to check it properly (e.g., using a calibrated thermometer probe placed inside the middle thickness of the cooked chicken)
3. Where to record the information (e.g., space to capture temperature)
4. What the expected result should be (e.g., equal or greater than 165 °F)
5. How to correct the out-of-compliance item (e.g., an app tells the user to recook the chicken for 1 minute and then recheck the temperature or to not use the cooking equipment and notify the owner/operator)
6. How to contact management when the corrective action cannot be performed, and to document the decisions made to ensure the safety of the food

Clearly, it would be difficult to manage all of these tasks using a paper-based checklist. Thus, digital technology enables more than a checklist, and can also actually reinforce training of the knowledge of each food safety control each time a manager performs an assessment.

Of course, checklists do have their value in a foodservice business; this is likely why they were originally used to monitor food safety controls such as food temperatures when this need was adopted within the industry. When a restaurant business requires its employees to prepare a certain amount of food ingredients to ensure enough food for a sales hour, they naturally use a checklist for employees to ensure they have prepped each item in the needed amount. Or, when a restaurant needs to count inventory to determine which items are in stock or need to be ordered, they use an inventory checklist based on a purchase order invoice. Likewise, if a restaurant business needs to ensure certain equipment is cleaned and sanitized properly at certain times during a shift, a checklist is an effective tool to ensure accomplishment of this task.

Technology is changing so rapidly that even a cursory review of devices and software options being used in restaurants for digital FSMS would quickly be dated in this book. However, there are three primary technologies currently being used for food safety management in the restaurant industry: mobile apps used on smartphones/tablets, IoT devices located in/on key equipment, and data analytics software. These will likely grow in use, especially as the new wireless/mobile standard, 5G, is implemented across the United States. Each of these is discussed in more detail below.

## Digital Food Safety Management Systems Better Facilitate Control of Hazards on a Daily Basis

Health inspections and third-party audits are necessary to assess a foodservice business's execution of food safety controls to prevent foodborne illnesses. The health inspection is especially necessary as a public health intervention to ensure accountability to execute food safety controls by a foodservice business serving the public. As you may recall from Chap.1, health inspections and third-party audits do not enable monitoring and management of food safety risks each day, nor do they foster daily accountability of the employee's actions by the responsible individuals (PIC/ Certified Food Protection Manager (CFPM)) in the food safety management process. In contrast to health inspections and third-party audits, digital FSMS assessments that monitor controls by the responsible individual (normally a CFPM) provide a manager with the management tool that is the best means to achieve Active Managerial Control (AMC) of food safety risk in a foodservice establishment. The most recent FDA risk study in restaurants (see Chaps. 3 and 4) demonstrated that those restaurants with a CFPM and who also were using a well-documented and established FSMS had the fewest number of foodborne illness critical violations related to the causes of foodborne illness (FDA 2018).

I believe that the most serious gap in restaurant food safety management is the absence of a CFPM managing food safety controls via (1) a FSMS based on a Process HACCP plan and Prerequisite Control Program and (2) using a digital management tool that enables the proper monitoring of controls, corrective actions, and daily documentation of AMC. In many food service businesses that I encounter, they most often have CFPMs working at each shift of the operations (as PICs) satisfying the regulatory requirement to have a CFPM on site, but rarely are they using a FSMS where they can directly apply and practice their knowledge in the direct management of food safety risk while they also focus on food quality and customer service. Evidence for this observation was demonstrated by the FDA where only 10% or less of the fast-food or full-service restaurants that were audited were found using an effective FSMS (FDA 2018). This is likely because many foodservice businesses only use paper checklist to evaluate some food safety controls like temperature of food, and they don't believe the manager has time to execute FSMS.

However, FSMS (e.g., preferably digitally based, with prescribed assessment of controls and corrective actions) actually empowers CFPMs to directly apply their knowledge to the daily operations of the foodservice business and provides them with a management tool to do so. The management methods of assessing the compliance to each food safety control continually reinforces the CFPM training and also reinforces and holds accountable the proper safe food preparation processes performed by employees. In my experience, based on food safety management performed within a multi-unit foodservice business with over 2000 restaurants, this naturally leads to continuous improvement of food safety SOP execution across the business (discussed in more detail below).

Digital FSMS can also help a multi-unit foodservice business track and utilize data from mulitiple location FSMS use in all its locations to take corrective actions with its food ingredient and product suppliers to the locations (source of safe food). When I was leading food safety program management for a multi-unit foodservice business, I would often discover supplier issues by reviewing restaurant daily self-assessment data (e.g., chicken not cooking completely due to thawing issues) from restaurants receiving ingredients from the same supplier. This is not normally feasible using data from paper checklist or only third-party audits of the same restaurants even when performed as often as every quarter. For example, digital data from FSMS use from multiple restaurants in the same day (each receiving frozen chicken from the same supplier) indicated that each restaurant noted in the assessment that an issue was found in cooking chicken properly (not reaching 165 °F after cooking the prepared chicken correctly). The chicken appeared to be cooked, but when cut open was found to be partially frozen. We discovered that the chicken supplier was distributing chicken to these restaurants that was too large to thaw properly under the defined thawing SOP (the product was consistently incorrect in size and weight specifications). These digital data enabled us to quickly investigate this issue and determine the root cause as chicken size and weight out of specifications; this enabled us to communicate the issue to all the restaurants using this supplier, remove the product from preparation, and require the supplier to take corrective action to return to specification compliance. Without the visibility to digital data, the cooking issue in each restaurant location likely would have been attributed to cooking platform issues, employee training, etc. and gone unnoticed until customers' complaints of undercooked chicken were made. In contrast, a paper checklist or third-party audit—performed on different dates, with data provided retrospectively—would likely never have provided this level of insight for fixing a supply chain issue that could have caused food safety risk, and was already causing food loss/waste.

## Recommended Components of Effective Digital Food Safety Management Systems

The primary components necessary to enable digital FSMS are:

1. The computer hardware/device (normally a mobile computer tablet/device or smartphone; my personal recommandation) capable of functioning in a restaurant environment with:

   - A connection to the Internet to enable data sharing, updates to the software, and communications to facilitate reporting and alerts (via exception reporting) to other stakeholders in real time via email, text, and phone calls
   - A means to ensure easy data entry including with Bluetooth IoT-connected devices like thermometers

- A means to store data on the device or to an Internet data storage server called a cloud

2. The computer software (normally the app or application) that can host the FSMS SOPs, requirements, corrective actions, and assessment functions (e.g., capture employee name to a sick log, provide notifications, etc.) to ensure data capture and analytics. Additionally, the software should minimally have these basic functions defined by the Process HACCP plan and the Prerequisite Control Program (see Chaps. 3 and 4):

- Assessment functions to monitor controls in the food preparation processes that will be executed daily:

  - Receiving/source of food (e.g., cold chain compliance of *Listeria monocytogenes* growth in RTE foods during transportation)
  - Storage (e.g., placement of raw chicken below RTE foods in cold storage to prevent cross-contamination)
  - Preparing (e.g., separation of raw chicken preparation from RTE chicken preparation)
  - Cooking (e.g., thawing chicken properly and heating at 165°F to kill *Campylobacter* on raw poultry)
  - Hot/cold holding (e.g., keeping cooked black beans at 135°F or higher to prevent the growth and toxin production of *Clostridium perfringens*).
  - Cooling (e.g., keeping RTE food below 41°F within 6 h to prevent the growth of pathogens in the food)
  - Reheating (e.g., reheating all RTE foods to 165°F to kill pathogens in cooled foods)
  - Serving (e.g., having customers use a new plate each time they visit the buffet to prevent *Staphylococcus* contamination of buffet foods)

- Assessment of personal hygiene controls (as part of the health policy; see Chap. 3) that include:

  - Wellness check and SOP for the proper restriction/exclusion of sick employees via use of a sick log

    Digital sick log tracks employee name and time and date of exclusion, including when the employee is allowed to return to work, removing the employee from the sick log after this time.
    Assessment process, including the ability to check sick log, to ensure no excluded employee is working in the facility before they are allowed to return to work.

  - Proper hand washing
  - Proper glove use
  - Personal appearance and hair restraint

- Predetermined corrective actions for each item evaluated—a process to ensure corrective actions are performed before any other item is evaluated and what to do if the corrective action fails

- Corrective actions should include the ability to identify the employee related to the compliance issue (e.g., not washing hands properly or at the appropriate time) and track repeat out-of-compliance behaviors to facilitate re-training if needed.

- Quick access to food safety information based on the most current FDA Food Code (e.g., how to calibrate a thermometer to ensure accurate temperatures are measured)
- A means to assign assessments to managers, schedule assessments, and track/report when assessments are not completed
- Cybersecurity where the data is stored that includes methods to delete incomplete or prior assessments and methods to detect false data entry by managers
- Ability to display data and perform analytics in a format that can enable root-cause determinations (e.g., to determine contributing factors caused by restaurant vs. supplier issue) and predictive analytics
- A means to switch between different FSMS assessments if different (e.g., Process HACCP FSMS vs. cleaning and sanitation FSMS)
- A means to accommodate new business models for assessment (e.g., home delivery, ghost kitchens, meal kits, etc.)
- A means to track customer complaints specific to the foodservice business location

The digital FSMA used by multi-unit foodservice businesses (restaurant chains) should also be scalable to a central dashboard/website to enable visibility to all locations' data, trend analysis of each Process HACCP and Prerequisite Control Program control, and web-based controls based on the business's specific SOPs and requirements.

# Digital IoT

The Internet of things is the future digital motor of modern food safety management. "Smart" sensors, thermometers, equipment, and communication devices connected to the Internet will significantly change the way we monitor controls of foodborne illness risk factors (see below). This is primarily because IoT devices can reduce human error in measurement and recording of information (e.g., temperatures) and enable more real-time monitoring of this and other important controls. Multiple IoT devices can also be connected to each other via the Internet so that data from each can be shared and integrated into FSMS (Fig. 7.1).

Currently, the more common IoT devices most often deployed in restaurants are wireless temperature sensor devices installed in refrigerators and freezers to monitor environmental temperatures using software to track, trend, and alert users when temperatures deviate from a set point. A more recent IoT device has been developed to enable Bluetooth active thermometers. When used to take internal temperatures

**Fig. 7.1** Example of IoT devices and how they may be integrated via computer hardware and software to collect, analyze, report, and store data related to monitoring controls in Food Safety Management Systems: (**a**) Internet connection for integration of data for the purposes of analytics, communications, reporting, and data storage, (**b**) RFID tags with temperature sensors on products in the supply chain, (**c**). computer hardware with software to manage the FSMS monitoring functions from a central location, (**d**) mobile devices (smartphone) with software to facilitate FSMS functions including assessments, (**e**) Bluetooth thermometer probes sensors that capture temperatures and wirelessly input data into the assessments performed with FSMS, and (**f**) environmental sensors that capture temperature and wirelessly input data into the assessments performed with FSMS (additional sensors to capture motion and door closures not shown)

of food, the thermometers transmit the temperature to a mobile app on a tablet/smartphone with software designed to log/record the temperature. Unfortunately, both of these technologies have their flaws related to wireless reception and interference, often leading to false alarms or inaccurate temperature measurements. With this in mind, a focus of new business ventures has been to develop more accurate and reliable sensors. Some technology businesses are innovating new devices and software to improve temperature monitoring of foods that enables a smartphone to capture surface temperatures of food products while also documenting the product via photo identification. The ZippyYum LLC GoTemp (see https://home.zippyyum.com/gotemp/) device captures the temperature and alerts the user when the temperature is above or below set point.

Another promising type of IoT device has been the temperature and location sensor devices that enable cold chain monitoring of products in the supply chain, end-to-end from supplier facility to distributor to restaurants. These devices have temperature sensors, radio-frequency identification (RFID), and GPS functions that can monitor temperature over time and report when the temperature of a specific product has exceeded set points. This enables employees at the time of receiving a delivery to reject products with temperature abuse before accepting the delivery, even if the current temperature is in compliance, based on the data from the product RFID device that may show prior temperatures out of compliance for too long.

RFID technology has also been deployed as IoT devices where products are labeled for date of expiration and location, enabling retail sales and foodservice businesses to track inventory, shelf-life, and FIFO to ensure safe source of foods. Other IoT devices that have been developed to monitor and report food safety controls include hand soap dispenser usage/counters to report numbers of hand wash events per hour, door closure sensors to monitor when and how long refrigerator and freezer doors are left open, and dish washer sanitizer/hot water sensors to alert the user that there is no sanitizer/hot water being used in the ware washing equipment. These devices are generally associated with equipment suppliers, and are not normally available for individual integration with current digital FSMS.

## Digital Data Analytics

The most important value of digital-enabled FSMS is in the data generated by the system and how ease of access to the data can be used to ensure food safety. These data can be used to trigger real-time alerts, recommend and confirm that corrective actions have been made, drive automatic equipment functions, provide data for tracking controls (e.g., who is on the sick log) during use to confirm controls are in place, and inform stakeholders when controls are absent to ensure prevention of hazards. Likewise, data analytics of the information from process HACCP assessments, combined with other food safety-related data, can be used to determine risk, enabling businesses to know where to escalate interventions at a restaurant. This might include immediate risk-based business-initiated third-party audits or field staff visits to the restaurant to immediately correct execution errors and prevent hazards.

For example, Chick-fil-A Inc. in partnership with Amazon (see The Science Times (2019) and AWS (2019)) has reported using data analytics to track customer complaints about their restaurant locations from social media data. Chick-fil-A developed a computer software analytics tool with Amazon that could reliably identify keywords, phrases, and customer tone from social media posts to help identify associated restaurants that might have an emerging foodborne illness risk. The Amazon AWS-hosted solution processes these data every 10 min from the most common social media platforms (searching words like "illness," "food poisoning," "vomit," "throw up," "barf," and "nausea"). AWS Comprehend, Amazon's natural language processing service, checks sentiment and determines legitimacy of the comments. This, in turn, enables the business to alert franchisee operators of a possible issue, which can then be tied to the recommended action(s) to reduce the risk. A senior principal IT leader of food safety and product quality at Chick-fil-A stated clearly that the intention is to use data analytics to help predict and address risk before they develop into a foodborne disease outbreak: "For us in this journey with analytics and food safety, we're going from a place of hindsight to insight … and eventually foresight so we can be more proactive in helping our Restaurants better identify and address food safety risks" (The Science Times 2019).

## A Case Study to Demonstrate the Efficacy of Digital Food Safety Management Systems to Prevent Norovirus Outbreaks

According to the most current data from the CDC on outbreaks in the United States, restaurants cause >60% of all outbreaks (CDC 2019). The number one agent/hazard causing foodborne disease outbreaks in the United States is Norovirus (which has only one species, *Norwalk*, named for a city where a large outbreak occurred in 1968). Interestingly, of all the restaurant outbreaks that occurred in 2017, Norovirus caused 46%, with 4092 illnesses, and a similar percentage year-to-year occurs according to the CDC. This is likely due to the fact that Norovirus is one of the most difficult hazards to prevent in foodservice: it can enter and then spread into the restaurant from multiple sources (from employees, customers, and foods); it is difficult to kill with sanitizers commonly used in foodservice; it is persistent on surfaces, where it can survive for several weeks; and it has a low minimum infectious dose required to cause illness. Further, people with Norovirus can shed the virus onto restaurant surfaces and food for weeks—even when asymptomatic with no signs of illness (Tung-Thompson et al. 2017).

Common employee-related issues and wellness demonstrate just how difficult it can be to manage just one of the controls for Norovirus prevention in foodservice:

- It is almost impossible to know when an employee is sick if they do not show signs of illness or report the symptoms to managers (vomiting and diarrhea), unless of course they physically vomit in the restaurant.
- Employees can be sick but have no symptoms (up to 2 weeks after being sick as well).
- Employees often work while infected (even though they know they have been sick) transmitting the virus from hands to environmental surfaces to food.
- Employees who are excluded from one restaurant often work at another restaurant (i.e., are employed by multiple businesses).
- Employees (discovered sick/excluded) will have already infected other employees who continue to work but don't yet have signs and/or symptoms of illness.

The CDC and FDA recommend several ways to prevent Norovirus outbreaks due to cross-contamination of foods in foodservice (Fig. 7.2), many of which are controls discussed as part of FSMS design in Chaps. 3 and 4. Because Norovirus can contaminate foods from multiple sources, multiple controls must be in place and monitored every day to prevent this one hazard (which can quickly lead to a foodborne disease outbreak) in a foodservice business. One only needs to Google a name of any large restaurant chain along with the word Norovirus to see how difficult these outbreaks are to prevent.

Imagine attempting to monitor each of these controls using a paper checklist. Alternatively, the best means for restaurants to ensure that each of the controls necessary to prevent Norovirus is in place can be accomplished using digital FSMS. For example, the most common cause of Norovirus entry into a restaurant is from work-

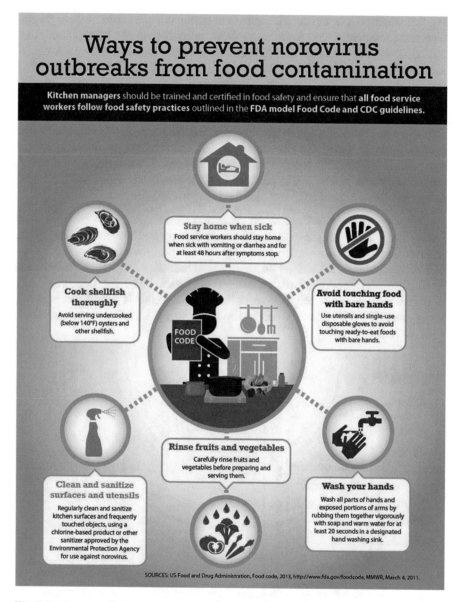

**Fig. 7.2** Recommended ways to prevent Norovirus foodborne illnesses and outbreaks in foodservice businesses by the FDA and CDC, demonstrating the complexities of multiple controls necessary to prevent this one hazard that are better monitored with digital FSMS. Each of the recommended controls should be included in the Prerequisite Controls Program and the Process HACCP plan according to the menu and monitored via FSMS on a daily basis. Additionally, the cleaning and sanitation FSMS should be monitored via an FSMS to ensure frequently touched surfaces are regularly cleaned and sanitized in addition to food contact surfaces and utensils with a focus on Norovirus prevention on high-touch point surfaces

ing sick employees (symptomatic), who then spread the virus onto surfaces (beginning most frequently with restroom surfaces, according to the FDA (Duret et al. 2017)). They also contaminate kitchen environmental surfaces (like cook equipment knobs/handles, refrigerator door handles, hand wash sink faucet handlers, etc.) and foods when they don't wash their hands properly and/or do not wear single-use food service gloves.

One of the most effective controls to prevent working sick employees (the most effective means to prevent an outbreak) is to establish a restaurant health policy (for the best example, see FDA's Employee and Personal Hygiene Handbook at https:// www.fda.gov/food/retail-food-industryregulatory-assistance-training/retail-food-protection-employee-health-and-personal-hygiene-handbook discussed in more details in Chaps. 3 and 4) that includes the following as priority:

- Train employees on when to report symptoms (fever, diarrhea, vomiting) and or diagnosed disease to the manager/PIC; inform them that they must comply with exclusion or restriction orders until they can show they are no longer risk.
- Managers should be trained on when to exclude sick employees (those with diarrhea and vomiting/fever or diagnosed illness) from the restaurant, and ensure they do not return to work at least for 72 h (this is what I recommend; however, the FDA Food Code states 24 h) using a digital wellness check (Fig. 7.3).

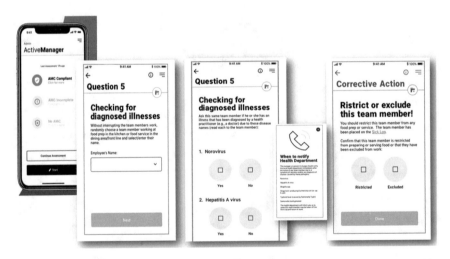

**Fig. 7.3** Example of digital FSMS using an app on a smartphone. A digital FSMS such as this enables the proper information for the CFPM to perform a detailed assessment of illnesses during a wellness check that is necessary to properly monitor and control the risk of working sick employees (questions/answers flow from left to right digitally as each answer is selected). According to the FDA Food Code, a manager must demonstrate knowledge of the required foodborne illnesses that they should screen and exclude employees from work or know if the employee or the health department reports these illnesses. Using digital FSMS ensures an accurate and consistent means to monitor controls in the Prerequisite Control Program and Process HACCP plan and provides and documents the recommended actions the CFPM should take

- Using a sick log to track excluded employees and ensure they do not return to work until no longer sick (Fig. 7.4).
- Although not all employees who are excluded from work will have had a Norovirus infection nor will those that recover from an illness shed virus when they feel/appear well (asymptomatic shedding), it is best to use a longer exclusion time due to these risks.

- Managers should do "wellness checks" for signs and symptoms before every employee "clocks" into work before each shift and during the day to reinforce to employees the need to report when they have symptoms of Norovirus.
- Ensure employees are washing their hands properly and at the correct times, especially after using the restroom (I recommend a double hand wash, one in the restroom and one before an employee is allowed to enter the kitchen).
- Ensure employees are using single-use foodservice gloves properly.

You can see that it would be difficult to establish a paper checklist to monitor, track, and manage each of these important controls including what to do after an employee is

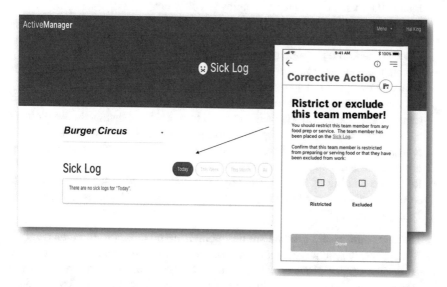

**Fig. 7.4** Example of digital FSMS using an app on a smartphone (foreground) and then managing the data via a desktop computer and software program (background). A digital FSMS such as this enables the proper information for the CFPM to perform a detailed assessment of illnesses during a wellness check on a smartphone, for example, that is necessary to properly monitor and control the risk of working sick employees (see Fig. 7.3). In this example, a CFPM may be reminded to determine if an employee should be restricted or excluded from work and, if excluded, placed on an employee sick log. The digital FSMS may then be used by this and other CFPMs to ensure the employee on the sick log is not working until removed from the log. Using digital FSMS ensures an accurate and consistent means to monitor controls in the Prerequisite Control Program and Process HACCP plan, and provides and documents the recommended actions the CFPM should take

excluded from work, just to prevent employee-related transmission of Norovirus to the restaurant environmental surfaces and cross-contamination of food. However, using a digital FSMS (in this example on a smartphone app, Fig. 7.3), multiple controls can be monitored based on where and when the hazard is likely to contaminate food. This would include monitoring proper cleaning and sanitation, excluding sick employees from work via a wellness check, and informing manager, when an employee is allowed to return to work after being excluded, and ensuring all personal hygiene controls necessary to prevent Norovirus food contamination are performed.

Thus a digital FSMS can first initiate the wellness check by notification (reminder) to a manager through the smartphone app; the manager can then ask the relevant questions (provided by the app) to each employee upon clocking into work or at different times during their shift, like asking all employees during a food preparation process. The app can provide the proper steps and actions to be performed by the manager via the following:

- Provide the manager with the proper questions to ask the employees in a wellness check.
- Enable the manager to select which employee is being interviewed.
- Enable the manager to enter an employee's answers.
- Inform the manager of what action is required for the employee based on the answers selected (including a legal requirement to call the health department based on certain reportable diseases).
- Enable the manager to track an excluded employee once the employee has been excluded by logging the employee's name into an electronic sick log (Fig. 7.4); the app might help alert different managers on other shifts to ensure an employee remains excluded until allowed to return to work by showing the employee's name on the sick log via the app or alerting managers that a restricted employee is working via IoT connection to a Point of Sales (POS) device that employees use to clock into and out of work.
- Reinforce training of the managers by teaching which diseases they should ask employees about according to the health policy including Hepatitis A, *Salmonella* Typhi, *E. coli* O157, etc.

The app could also a provide similar monitoring processes to the managers for each employee's practice of personal hygiene controls that include correct hand washing and proper glove use (King and Michaels 2019). Because individual employees would be monitored during their shift, each could be required to be retrained on the controls they are responsible for as a corrective action when multiple compliance issues are tracked and reported by the digital FSMS. If such a digital FSMS was in place and used by every CFPM at each shift during operations within a foodservice business, it could have a significant impact on the prevention of Norovirus illness and outbreaks. This is because it would enable continuous monitoring of each employee for signs/symptoms of illness but also reinforce the training of both employees and managers on the health policy (due to continuous dialog), which requires both to report when they are sick BEFORE they report to work.

Second, digital FSMS could inform the manager to check the sanitizer in use, its concentration (Fig. 7.5), and also its proper use at the proper times in the foodservice environment to reduce the spread of Norovirus in the foodservice facility that may occur due to transmission of Norovirus to high-touch surfaces in the facility (Duret et al. 2017). Because asymptomatic employees and those that may not report when they are sick may also work in the restaurant, it is important to ensure that a sanitizer is used at the proper strength and used to regularly clean and sanitize these and other surfaces in the restaurant. Because an employee working sick is the most likely event leading to foodborne illnesses caused by foodservice businesses, a digital FSMS may provide for a means to initiate additional alerts to managers when an employee is found working sick or "calls in" sick and track when customers report illness in the same time period, perhaps via data analytics in the software that captures the two events and then triggers an alert/notification to all managers/owner to escalate cleaning and sanitation frequency to reduce Norovirus on surfaces as discussed in Chap. 4.

Of course, the other controls (Fig. 7.2) of Norovirus in a foodservice business are important as well, but are not as high risk as working sick employees. These include ensuring produce is washed properly, foods are cooked to the proper temperature (especially shellfish), and any body fluid spills (e.g., a customer or employee vomit-

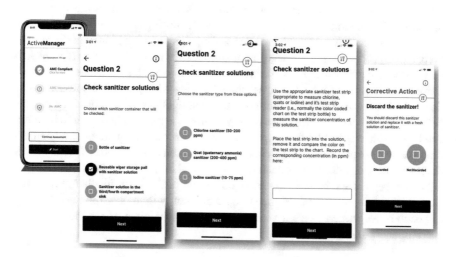

**Fig. 7.5** Example of digital FSMS using an app on a smartphone. A digital FSMS to monitor proper cleaning and sanitation in a foodservice facility such as this enables the proper information for the CFPM to perform a detailed assessment of proper sanitizer concentration during a cleaning and sanitation check on a smartphone. In this example, a CFPM may be reminded to determine what sanitizer type is in use and then to check the sanitizer and record results during the assessment (questions/answers flow from left to right digitally as each answer is selected). Using digital FSMS ensures an accurate and consistent means to monitor controls in the Prerequisite Control Program and Process HACCP plan, and provides and documents the recommended actions the CFPM should take

ing in the restaurant) are properly cleaned and disinfected as discussed in Chap. 4 using a FSMS. Each of these can be evaluated and managed within digital FSMS.

## The Future of Digital Food Safety Managements

The expansion of technologies developed for foodservice FSMS is growing, and new IoT devices will continue to invade the restaurant space as the 5G mobile network is implemented across the United States. Mobile apps will continue to be developed to enable food safety management tools and software to enable the design and the implementation of FSMS for any business with a smartphone. The future food safety data management and analytics will include machine-learning risk assessments to alert (predict) pending risk before they occur based on historic data from multiple sources in real time. Machine-learning epidemiology has already been developed between Google and the Harvard T.H. Chan School of Public Health where a model, called FINDER, was used to detect foodborne illnesses from restaurants in real time in two large cities (Sadilek et al. 2018). Using anonymous aggregated web search and location data of persons who visited a particular restaurant and later searched for terms indicative of food poisoning, the researchers were able to identify potentially unsafe restaurants subsequently confirmed by a foodborne illness risk factor health inspection. Some foodservice businesses are innovating digital FSMS technology themselves, for example, testing computer vision systems using IoT cameras that "watch" employees for proper actions as they perform food preparation processes. For instance, employees who have been handling raw chicken may be "watched" by a camera that then alerts them to wash their hands before they move to other areas of the kitchen. Machine learning will also inform more automation in FSMS, including smart thermometers that tell the refrigerator to lower temperatures to cool foods down more rapidly or smart sensors embedded into holding/ hotel pans that tell hot-holding equipment to heat products up when temperatures drop below a set point temperature. I can also envision that, someday, digital FSMS data from these devices may be shared with the local health department and, together with local regulators, be used to ensure the public health from foodborne illnesses caused by foodservice establishments more efficiently than only relying on health inspections performed twice a year or customer complaints of foodborne illness (see Appendix B).

## References

Amazon Web Services (2019) Chick-fil-A uses Amazon's Comprehend to help spot foodborne illness. See: https://aws.amazon.com/solutions/case-studies/chick-fil-a/

Centers for Disease Control and Prevention (2019) Surveillance for foodborne disease outbreaks, United States, 2017. Annual report. U.S. Department of Health and Human Services, CDC, Atlanta

Duret et al (2017) Quantitative risk assessment of norovirus transmission in food establishments: evaluating the impact of intervention strategies and food employee behavior on the risk associated with norovirus in foods. Risk Anal 37:1–27

Food and Drug Administration (FDA) (2018) FDA report on the occurrence of foodborne illness risk factors in fast food and full-service restaurants, 2013–2014. See: https://www.fda.gov/food/cfsan-constituent-updates/fda-releases-report-occurrence-foodborne-illness-risk-factors-fast-food-and-full-service-restaurants

King CH, Michaels B (2019) The need for a glove-use management system in retail foodservice. Food Safety Mag. June/July. See: https://www.foodsafetymagazine.com/magazine-archive1/junejuly-2019/the-need-for-a-glove-use-management-system-in-retail-foodservice/

Sadilek A et al (2018) Machine-learned epidemiology: real-time detection of foodborne illness at scale. NPJ Digit Med 36:1–7. See: https://www.nature.com/articles/s41746-018-0045-1

The Science Times (2019) Chick-fil-A's AI can spot signs of foodborne illness from social media posts with 78% accuracy. See: https://www.sciencetimes.com/articles/22143/20190527/chick-fil-a-s-ai-can-spot-signs-of-foodborne-illness-from-social-media-posts-with-78-accuracy.htm

Tung-Thompson G, Escudero-Abarca BI, Outlaw J, Ganee A, Cassar S, Mablat C, Jaykus L-A (2017) Evaluation of a surface sampling method for recovery of human noroviruses prior to detection using reverse transcription quantitative PCR. J Food Protect 80:231–236

# Chapter 8
# The Business Value Proposition in Using Food Safety Management Systems

Every business requires a value proposition or return on investment to justify spending, especially for non-direct sales-related expenditures. For example, providing sanitizing hand wipes for children at a restaurant play area where children play and eat doesn't immediately induce a purchase intent (sales) by a customer. Instead, the value proposition to the business is customers' perception that the restaurant is clean and cares about children's welfare not just selling them food. This can lead to customers visiting the facility more often with their children and thereby increase purchasing intent. Food Safety Management Systems (FSMS) should not necessarily require a value proposition: in my personal opinion they should be required of all restaurants (by regulatory authorities), expected by customers (good business), and should be the priority of a foodservice business (to stay in business). However, as discussed in Chap. 2, many foodservice businesses have not experienced a foodborne disease outbreak and may not be aware that they are causing sporadic cases of illness, which can lead to complacency until an issue occurs. Thus, by showing the value proposition or return on investment in using FSMS to a foodservice business beyond preventing foodborne illnesses (which is the most Public Health value of course), the business can also build the FSMS cost into its business model.

## Prevent Foodborne Illnesses for the Benefit of the Public Health

It is difficult to quantify how many foodborne illnesses and outbreaks are prevented by FSMS in restaurants. However, data demonstrate a quantified relationship between the observance of critical foodborne illness risk factors (also called critical food safety violations on a health inspection report) and causation of foodborne illnesses and outbreaks (as discussed in Chap.2; see also Petran et al. 2012). The FDA performed a 10-year study to investigate the relationship between FSMS under

© Springer Nature Switzerland AG 2020
H. King, *Food Safety Management Systems*, Food Microbiology and Food Safety, https://doi.org/10.1007/978-3-030-44735-9_8

active management by a Certified Food Protection Manager (CFPM) and the occurrence of risk factors and food safety behaviors/practices commonly associated with foodborne illness in restaurants (FDA 2018). The FDA analyzed hundreds of restaurants across the United States, representing a significant sample of the proportion of the US population that patronize restaurants. Investigators performed an unannounced, non-regulatory audit using the FDA Food Code to inspect 396 full-service restaurants and 425 fast-food restaurants in multiple states. The number one factor that correlated to fewer foodborne illness prevention compliance issues in both fast-food and full-service restaurants was the execution of a daily FSMS that could achieve Active Managerial Control of the foodborne illness risk factors. The presence of CFPM also correlated with restaurants having a more effective FSMS as long as the manager was the person-in-charge (i.e., was present at the time of the inspection); however, a CFPM's impact did not replace the benefits of FSMS when these were absent during operations. Interestingly, multi-unit-associated restaurant business locations (i.e., franchisees or restaurants owned by a restaurant chain) had significantly lower instances of out-of-compliance items compared to independent single-unit businesses; likely due to more well-developed and established FSMS developed by food safety experts on staff.

The presence or absence of actual microbiological hazards (the most common and more directly correlated to the risk of a foodborne illness) is also reduced when FSMS that provide Active Managerial Control of the foodborne illness risk factors are executed in a foodservice business. A study by Lahou et al. found that, during a period of 1 year, when a new FSMS (e.g., focused on monitoring cleaning and sanitation) was implemented in a restaurant operation, there was a sustained compliance to microbial quality metrics that included lower prevalence of viable pathogens (*Listeria monocytogenes*, *Salmonella*, *Bacillus cereus*, and *Escherichia coli* O157) on restaurant food contact and non-food contact surfaces (Lahou et al. 2012). Two of the associated factors for this improved compliance were proper supplier selection and certification (according to microbiological specifications and auditing to verify compliance of suppliers when food was delivered to the restaurants), which reduced the numbers of pathogens on incoming raw foods. Pathogens on raw animal meats and poultry are a significant source of cross-contamination to foods from employees' hands and via misusing gloves (see King and Michaels 2019).

## Prevent Associated Cost of Foodborne Illness Mitigation

The restaurant industry is a multibillion-dollar (>$863 billion in 2019) annual economic value in the United States, with the average citizen eating out about five times per week (National Restaurant Association 2019). With >300 million persons in the United States, restaurants are a staple food source for >60 million people; the impact on their health is significant when the food is not safe. The USDA has reported that, overall, a conservative approximation of all foodborne illnesses caused over $15.5 billion in economic loss in 2014 US dollars, primarily when measured by medical cost and productivity losses (Hoffmann et al. 2015). Estimates of the incidence of foodborne disease acquired in the United States, and therefore economic burden estimates, are often very uncertain. The US Centers for Disease Control and Prevention (CDC) estimates that the foodborne disease incidence from just 15 pathogens (reviewed in Hoffmann et al. 2015) could range from 4.6 million to 15.5 million cases in a typical year. These 15 pathogens cause 95% or more of the foodborne illnesses, hospitalizations, and deaths in the United States in which a specific pathogen can be identified. These 15 pathogens include *Campylobacter* spp., *Clostridium perfringens*, *Cryptosporidium* spp., *Cyclospora cayetanensis*, *Listeria monocytogenes*, Norovirus, *Salmonella* non-typhoidal species, *Shigella* spp., STEC O157, STEC non-O157, *Toxoplasma gondii*, *Vibrio vulnificus*, *Vibrio parahaemolyticus*, *Vibrio* other non-cholera species, and *Yersinia enterocolitica*. Based on this range of incidence estimates, economic burden has been estimated to range from $4.8B to $36.6B in 2013 dollars (Hoffmann et al. 2015). Many of these illnesses are known to occur due to foodservice businesses. With approximately 60% of the annual foodborne illnesses caused by foodservice businesses, the economic burden to the United States is likely significant on an annual basis.

These costs to the US economy should be sufficient incentive for the foodservice industry and regulatory agencies to work together to prevent foodborne illness and outbreaks. However, because most businesses are local, are driven by profit/loss business principles, and not all experience a national foodborne disease outbreak (or realize the true degree of sporadic cases they may cause; see Chap. 2), it is equally important to show the business cost (lost revenue, employee labor, cost of lawsuits and legal fees, fines, increase in risk management insurance premiums) associated with foodborne illnesses.

Several organizations have reported the estimated cost of customer-associated foodborne illnesses to the restaurant industry in the past, primarily based on aggregate annual cost to individuals or society at large. A recent computational model published by researchers at the John Hopkins Bloomberg School of Public Health provided a more useful model for the predicted cost of a confirmed foodborne illness outbreak (just two or more persons) to a single restaurant location. The model is based on the cost after a single foodborne disease outbreak (Bartsch et al. 2018) using actual cost from published data from the CDC, the Bureau of Labor Statistics, and other relevant restaurant industry and regulatory sources related to the probabilities of events that drive cost (e.g., of hospitalization of persons ill with *Salmonella* infection, hourly wages lost, cost of meals lost, cost of training, etc.). The model included these inputs (see below) to predict cost after an outbreak of between 5 and 250 persons infected with 1 of 15 foodborne pathogens that were the cause of prior restaurant-associated outbreaks to a fast-food restaurant, a fast-casual restaurant, a casual-dining restaurant, and a fine-dining restaurant:

- Total productivity losses for ill restaurant employees' cost
- State and local health department inspection fees and fines
- Hospitalization and mortality cost
- Customer lawsuits and legal fees
- Lost revenue from lost customer count cost (sales measured by decreased meals sold per illness)

- Employee re-training cost
- Risk management insurance premium increases

The model also demonstrated the predicted cost of an outbreak to a restaurant comparing two of the most common causes of foodborne illness outbreaks in the United States, *Salmonella* and Norovirus. A single outbreak of 250 persons due to *Salmonella* for 1 restaurant location was predicted to cost up to $2,075,561, while a Norovirus outbreak under the same conditions with 250 persons was predicted to be $2,058,188 (both using probabilities of hospitalization and mortality data among other variables discussed above derived from the CDC). The cost of lawsuits and legal fees and the number of ill people in an outbreak were the biggest drivers of cost to a single restaurant. The median cost of an outbreak at a fast-casual restaurant was $979,793—without costs from lawsuits and legal fees—but went up to $1.81 million with those costs. Finally, and more importantly, the model showed that one outbreak in just one restaurant location could consume a substantial proportion of a fast-food restaurant's annual profit. For example, a single 15-person foodborne ill-ness outbreak could consume between 0.2% and 68.3% of a single fast-food restau-rant's average annual sales (of $2.6 million, Jones 2015), 0.3% and 72.3% of a fast-casual restaurant location's annual sales (of $2.47 million, Tice 2014), and 0.5% and 101.1% of $1.8 million average annual sales of just one franchisee restau-rant in the top 100 independent restaurants in 2015 (using data reported in the pub-lication, Restaurant Business 2016).

Additional costs of mitigation are well known and would be experienced at the corporate business level in a multi-unit restaurant chain. One could predict to incur additive cost when the number of restaurants involved in the outbreak is multiplied (one vs. 100)—where, for example, each received and served adulterated ingredients or products across several states (e.g., if associated with a romaine lettuce outbreak caused by *E. coli* O157 due to the unsafe source of food risk factor). Recalls, removal of food from facilities, employee/customer vaccinations (e.g., if customers were exposed to Hepatitis A from a food), cleaning services (Norovirus-related restaurant closures and remediation to disinfect the restaurant using professional services), employee replacement (recruitment and onboarding), and PR/marketing costs to recover customer trust of the brand are all additive costs to the foodservice business.

The expected cost to a restaurant to implement FSMS to reduce the risk of food-borne illnesses and outbreaks is significantly less than the cost of causing them. As an example, the cost of these minimum components of an FSMS for a single food-service business might include:

- $1,178.85 for three Certified Food Protection Managers (CFPMs) required to work three shifts and manage the FSMS—at a cost of a $152.95 online training course per manager (for estimated cost per ServSafe Manager course, see https://www.servsafe.com/access/ss/Catalog/ProductList/10) and estimated labor hours of 16 h at $15 minimum wage/h per manager
- $375 for 25 online food handler courses so that each employee has knowledge of the risk they can cause from working while sick, not wearing gloves properly, etc.—at a cost of $15 per employee

- $600 per year for one mobile food safety management system app—at a cost of $50/month per location (that can be used by each manager's smartphone)

Comparing this estimated total cost—$2,153.85 to implement FSMS for 1 year (and approximately $1000 per year thereafter to maintain due to likely turnover of employees)—to the lowest foodborne illness outbreak cost modeled by the John Hopkins researchers (Bartsch et al. 2018) from just one foodborne disease outbreak of only five persons in a fast-food restaurant ($3,968), a fast-casual restaurant ($6,330), a casual-dining restaurant ($8,030), and a fine-dining restaurant ($8,273) with no added lost revenue, lawsuits, legal fees, or fines included in this cost demonstrates that prevention of a foodborne disease outbreak is far less costly than actually waiting until one is likely to occur not using FSMS.


## Prevent Health Inspection Violations and Improve Health Inspection Scores/Grades

Implementing proactive FSMS can work to prevent poor health inspection scores/grades by ensuring food safety controls are active at all times including during a health inspection. When a foodservice establishment performs a daily assessment via use of its FSMS, (1) the hazards are controlled in a more uniformly and timely manner (via immediate corrective action by the business), (2) the controls are most likely to be sustained over time, and (3) it will enhance the value of the regulator-operator relationship, resulting in better health inspection scores/grades because violations are less likely to occur. This also likely reduces the need for additional follow-up inspections by regulators, enabling them to focus on higher-risk restaurants that do not use FSMS. Supporting evidence for this has been reported by state regulators, using announced inspections as part of an Active Managerial Control (AMC) education program for restaurants in Minnesota. After the initial education program was implemented in participating restaurants, critical violations of foodborne illness risk factors were reduced by one-half compared to prior inspections data (Reske et al. 2007). Likewise, the FDA foodborne risk factors study described throughout this book (FDA 2018) showed that during inspections, the average number of foodborne illness risk factors was significantly lower in all restaurants that had a well-developed, observable FSMS in place compared to those that had no observable FSMS (see FDA Summary, Fig. 8.1).

Social media and online platforms (e.g., Google, Yelp, Twitter, etc.) that enable customers to post information about businesses and products continue to be an important economic driver in the restaurant industry; negative statements related to a customer's experience(s) with a restaurant's food can lead to loss of sales for the business. This is equally true of regulators' inspection reports (e.g., via an official health inspection score/form posted to these platforms). A recent study reported by researchers at Harvard Business School and Lehigh University was performed to determine how posting of health department scores from restaurants in the city of

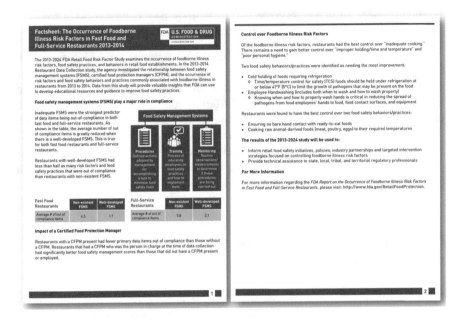

**Fig. 8.1** Evidence for the major role FSMS play in the improvement of compliance to foodborne illness risk factors measured by the FDA, thus improving health inspection scores/grades by food-service businesses. Factsheet from the 2018 risk factors study. (Source: FDA 2018)

San Francisco on the online platform Yelp (Dai and Luca 2019) might affect customer restaurant demand. The customer restaurant demand was measured as an indirect purchase intention (not actual sales) and pulled from the customers' engagement with a specific restaurant on Yelp. The investigators found that, of all restaurant scores posted, low scores led to a 13% decrease in purchase intentions for a restaurant. Interestingly, when a "hygiene alert" was added that appeared only on Yelp restaurant review pages that had low health inspection scores (Yelp tested this alert as an additional means to highlight the risk), there was a 20% total decrease in purchase intentions. Clearly, improvement of health inspection data can help sustain and even grow sales in the restaurant.

## Enable Foodservice Operators to Address False Claims from Customers

Before I discuss this subject matter, let me first provide a personal opinion. Any restaurant business that causes a confirmed foodborne illness or injury of any type (biological, chemical, or physical) should be required to and be prepared to pay the cost of this illness including medical cost to a customer. When a person or persons are harmed or injured by a business's food, the business owners must do the right thing and take responsibility for the issue; the business also protects its reputation by demonstrating integrity. Of course, in nearly all cases, foodborne illnesses are not intentionally caused by a restaurant. However, the fact that all restaurant business owners can know how to prevent them (based on health inspection reports of their business that show which foodborne illness risk factors are not in compliance, available food safety education resources they can use to know how to prevent foodborne illnesses and outbreaks and train their employees, and numerous vendors that sell effective food safety management tools) *but they choose not to do so* supports an intention to allow this risk and thus, in my opinion, could support a foodborne illness claim by customers.

However, when a foodservice business owner has committed to implementing FSMS that the health department has validated, the business must be prepared to address the many false claims of foodborne illness that will occur (in which there is little evidence to support the causation of the illness and where the restaurant is clearly executing FSMS to monitor and control foodborne illness risk). A false claim could be defined as a complaint of injury or illness that cannot be substantiated by evidence such as specific purchase history for date/time of purchase, information about the food item from the menu consumed by the person, information of when the person started to get sick and with what symptoms, and when the time to illness after consumption is outside of the average incubation period necessary to cause the illness. On the other hand, a quick judgment to disregard or deny a legitimate foodborne illness claim could also be very damaging to a restaurant business and to the public health. The most supportive evidence for the cause of a foodborne disease illness (see Fig. 8.2) is the "match" between a hazard in the food (most often a pathogen, but could also be an allergen or physical hazard such as bones in chicken nuggets), combined with evidence that the customer purchased the suspect food (via purchase and traceback history), and a medical practitioner's diagnosis (with or without a clinical specimen of the customer, usually a fecal sample) within a defined window of time from a certified laboratory.

The recent adoption of more precise technologies such as whole-genome sequencing (WGS) and supply chain traceback analytics—used by the CDC and many state laboratories to more equitably match a hazard to a diagnosis with a clinical specimen from a customer—provides all the evidence needed to support a claim against a foodservice business. However, because almost all legal claims against a foodservice business are normally civil cases (rather than criminal cases), circumstantial evidence such as clinical diagnosis of a foodborne illness—even with no clinical specimen matched to a window in time (normally the average incubation period of a disease) and purchase history of a food—can be enough to encourage further pursuit of a legal claim against a restaurant. Even though these claims may also be valid and should be thoroughly investigated (especially when there is more than one customer complaint/claim, which suggests an outbreak), there are likely many more claims against a restaurant business that are not valid.

William "Bill" Marler, of Marler Clark, LLP, is a well-known food safety lawyer who has been successfully investigating and litigating foodborne illness cases for nearly three decades. Mr. Clark believes that the food industry, from farmer to

**Fig. 8.2** CDC infographic showing how foodborne disease outbreaks are performed to link a foodborne illness to the source of contaminated food. (Source: CDC, www.cdc.gov/foodsafety/outbreaks)

retailer/restaurants, tends to over-emphasize the many false foodborne illness claims (and, in my opinion, become complacent as a result), thus under-valuing the legitimate claims. Because his firm is contacted "virtually every day" by people who have become sick (usually only a few hours after consumption) and believe that their illness resulted from consumption of a particular food item, Mr. Clark uses and has published his quick, reliable method of distinguishing between a legitimate foodborne illness claim and a suspect one (Marler 2011). The vast majority of claims to his firm, he states, do not make it through this initial screening process as probable valid claims. However, the following types of evidence he uses to specifically tie a customer claim to a restaurant location suggest the validity of a claim:

- A health department investigation of an outbreak or incident of the restaurant
- Prior health department inspections data of the restaurant showing the presence of one or more foodborne illness risk factors
- Medical records of the customer's illness
- Lab results of the customer's illness

Marla shows that the stronger the medical and lab result records for diagnosis of a foodborne illness combined with food purchase history from the restaurant, the more important health department data supports the likely source of exposure as being from that restaurant.

The health department data that will support a foodborne illness claim can be numerous, do not require a poor overall score or grade, and include all of the CDC contributing factors (see Chap. 2) to the top five foodborne disease risk factors. For example, a restaurant can receive a 90%/A health inspection score/grade on its most recent inspection report with only one critical violation. But if the critical violation is correlated to a foodborne illness risk factor (an event known to cause a foodborne illness, e.g., employee working while sick at similar time when customer claims illness with vomiting and diarrhea), this health department data would support a foodborne illness claim related to this risk factor. In contrast to biological hazards like a Norovirus illness, most contributing factors of allergen or physical hazard injuries do not necessarily require medical records or lab results; thus, validity is more strongly tied to health department investigation and/or source of food evidence (a risk factor) matched to a customer's proof of purchase history. In other words, when a foodborne illness risk factor is documented in a foodservice business by a health inspection and two or more customers file a claim related to this same risk factor during the same time period, evidence shows, and the CDC reports that these foodborne illnesses are more probable.

Let's briefly revisit our case study in Chap. 1 (see "A Case-Study: One Incident, Two Different Outcomes") to further demonstrate this point. In this case study, a customer makes a claim of a foodborne illness, which may be a part of an outbreak that is likely due to Norovirus cross-contamination in a restaurant. The health inspector then visits the restaurant to investigate the validity of the claim. In this scenario, if the health inspector observes that the manager:

- Allows an employee to work while likely sick, as evidenced by frequent visits to the restroom, without inquiring whether the employee is sick
- Does not require glove use when handling RTE foods
- Does not have a verified hand washing policy in place
- Does not ensure that foods prepared by employees are cooked properly
- Does not ensure that sources of foods such as produce are safe sources

there is a likely foodborne disease outbreak occurring in the community (which of course likely initiated the health inspection). The lack of these controls dramatically increases the likelihood that the foodborne illness claim is valid. Although this scenario is fictional, it is based on actual experiences and observations. Evidence of food safety controls via FSMS that result in AMC of foodborne illness risk factors would reduce the probability of both cause and association. Thus, when a customer first calls a foodservice business or the health department calls the business in response to a customer's foodborne illness claim, the knowledge of and evidence that these controls are in place can inform the response to this initial complaint, and can actually prevent the need for further investigation or a legal claim.

One of the most important business functions in a foodservice business related to customer complaints is to resolve the complaint and ensure the person making the complaint remains a customer (sometimes called customer recovery). It becomes easier to respond to a customer complaint (remember to not assume it is false outright) when FSMS are in place that monitor and ensure/document that all of the proper controls of the foodborne illness risk factors were in place during each day of foodservice operations. With this information and data from a Process HACCP-based FSMS, there is a higher probability of customer recovery. As an example, a foodservice business owner/operator could use these steps in response to a customer's complaint of illness, but only if there are no other complaints during the same time period or about the same food item:

1. If feasible, check your FSMS data from the time/date of purchase for the customer's complaint before responding to the customer.
2. Ask about the customer experience that led to this complaint:

    (a) Obtain customer name, address, email, and phone number if possible.
    (b) Ask what they ate from your restaurant and when (and/or request purchase history). Also establish when the food item was consumed: immediately after purchase or later. If it was consumed later, inquire how was it stored prior to eating.
    (c) Ask when (date and time) did the customer actually get sick and what were the symptoms (listen for symptoms like diarrhea, vomiting, fever, fatigue, etc.). However, do not state these symptoms to avoid leading the customer.
    (d) Ask whether the customer went to see a medical practitioner or doctor. You are not permitted to ask what the diagnosis was or lead the customer to disclose this information (see HIPAA law here: https://www.hhs.gov/hipaa/for-individuals/guidance-materials-for-consumers/index.html), however, you may capture this information if offered. There are exceptions for asking employees about their illness found here relevant to wellness checks: https://

www.cdc.gov/nceh/ehs/activities/can-restaurant-managers-talk-with-sick-workers.html

3. Express sympathy without accepting blame/cause:

   (a) Ask how they are feeling now and whether they have started feeling better and when.
   (b) Express that we all have had foodborne illnesses and nobody expects them or likes going through one. Add that you appreciate their complaint just in case there is a risk that could help prevent others from getting sick from any restaurant.

4. Express the importance of and your dedication to food safety management in your restaurant:

   (a) Describe why it is important to you, e.g., your children or grandchildren eat in your restaurant or you know many children do, etc.
   (b) If you have not had any other customer complaints within the last 7 days, mention this.
   (c) Mention the primary controls you have in place such as requiring no bare-hand contact with RTE foods.
   (d) Describe how you checked the data (or will check; see bullet 1e above) from the FSMS you use to prevent risk on the day of the customer's purchase. If you found that food safety controls were in place (remember that the controls most commonly associated with an illness are the top five risk factors we have discussed numerous times throughout this book).

5. Conclude your conversation with the customer:

   (a) Thank the customer for informing you of the complaint.
   (b) Let the customer know you will continue to focus on food safety management for their safety.
   (c) You may want to offer the customer a free meal or drink to encourage them to return as a customer and to show your confidence in your food safety management program. However, be careful to not associate the offer of a free meal as a replacement for the prior meal but just more evidence of your confidence of your food safety management for their safety.

This is by no means a legal or risk management/insurance claim investigation procedure; it is intended as a best practices example based on my experience of measures that will assure a customer of food safety when actual FSMS are active and documented. It can also reduce the need for false customer complaints to be pursued further as legal claims.

When there are two or more claims of foodborne illness from different customers that ate the same menu item purchased on the same day, this would not be the time to focus on customer recovery. The best method for a more formal investigation of two or more foodborne illness complaints is to use a resource (developed for the retail foodservice and sales industry in conjunction with state regulators and the food industry) by the Council to Improve Foodborne Outbreak Response (CIFOR; see https://cifor.us) called the *CIFOR Foodborne Illness Response Guidelines for Owners, Operators, and Managers of Food Establishments* (see https://cifor.us/ clearinghouse/cifor-foodborne-illness-response-guidelines-for-owners-operators-and-managers-of-food-establishments). I highly recommend this resource, especially in preparation before an event. Many of the tools used in this resource will be easily found in your Process HACCP plan and your Prerequisite Control Program. Using this resource can also help you assist regulators in finding the link to a foodborne illness outbreak more quickly, which may prevent further losses to your business, and more importantly identify the cause and implement appropriate corrective actions before others are sickened. Some of the tools in this resource include (which are also useful to prepare your process HACCP plan):

- Illness Complaint Tracking Log Form
- Customer Foodborne Illness Complaint Form
- Menu Ingredient Listing Form
- Example of Menu Ingredient Listing
- Employee Communications Meeting
- Employee List Form
- Employee Health Assessment Form
- Employee Illness Decision Guide
- Example of Employee Illness Decision Quick Guide for Person-in-Charge (for Non-highly Susceptible Population Establishments)
- Product Sampling Procedure
- Sample Chain of Custody Form
- Distributor and Supplier Information Form
- Sample Food Establishment Food Safety Checklist Form
- Food Flow Chart Form
- Example of Employee Workstation Schematic
- Sample Re-opening Self-Inspection Checklist Form

The importance of brand protection to the success and even survival of a foodservice business in the United States cannot be overstated to anyone that understands how quickly a foodservice business can be put out of business due to a foodborne disease outbreak. A restaurant sustains and grows its business by the outcome of how a customer experiences their food and the restaurant environment; food quality and restaurant cleanliness have always been important drivers for growing sales in the foodservice industry. There are likely more benefits to using FSMS in a foodservice business that include lower food waste/cost, better customer service from trained employees, lower equipment maintenance cost, fewer pest infestations, and lower insurance cost due to effective mitigation of false foodborne illness claims, etc. However, many of documented evidence for these values cannot be reported (but have been experienced) because the data remains in protected corporate documents. Nevertheless, the value proposition for implementing FSMS is overwhelming, and the hope is that, when each and every foodservice business in the United States uses FSMS because they become part of their business model, there will be significantly fewer foodborne illnesses and outbreaks caused by the foodservice industry.

# References

Bartsch S et al (2018) Estimated cost to a restaurant of a foodborne illness outbreak. Public Health Rep 133:274–286

Dai W, Luca M (2019) Digitizing disclosures: The case of restaurant hygiene scores. Working paper 18–088. Harvard Business School. Electronic copy available at: https://ssrn.com/abstract=3131900

Food and Drug Administration (FDA) (2018) FDA report on the occurrence of foodborne illness risk factors in fast food and full-service restaurants 2013–2014, FDA National Retail Food Team

Hoffmann S, Maculloch B, Batz M (2015) Economic burden of major foodborne illnesses acquired in the United States, EIB-140, U.S. Department of Agriculture, Economic Research Service

Jones F (2015) Top 10 most profitable restaurant franchises in the United States. Food, Drink & Franchise. http://www.fdfworld.com/top10/495/Top-10-Most-Profitable-Restaurant-Franchises-in-the-United-States

King H, Michaels B (2019) The need for a glove-use management system in retail foodservice. Food Safety Mag. June/July. https://www.foodsafetymagazine.com/magazine-archive1/junejuly-2019/the-need-for-a-glove-use-management-system-in-retail-foodservice/

Lahou E, Jacxsens L, Daelman J, Van Landeghem F, Uyttendaele M (2012) Microbiological performance of FSMS in a restaurant's operations. J Food Prot 75:706–716

Marler WB (2011) Separating the wheat from the chaff. Marler Blog. Electronic copy available here: https://www.marlerblog.com/lawyer-oped/separating-the-wheat-from-the-chaff-the-reality-of-proving-a-foodborne-illness-case/

National Restaurant Association (2019) State of the industry. National Restaurant Association, Washington, DC

Petran RL, White BW, Hedberg CW (2012) Health department inspection criteria more likely to be associated with outbreak restaurants in Minnesota. J Food Prot 75:2007–2015

Reske KA, Jenkins T, Fewrnandez C, VanAmber D, Hedberg GW (2007) Beneficial effects of implementing an announced restaurant inspection program. J Enivron Health 69:27–34

Restaurant Business (2016) Top 100 independents: the ranking. http://www.restaurantbusinessonline.com/special-reports/top-100-independents

Tice C (2014) Seven fast-food restaurant chains that rake in $2M per store. Forbes. https://www.forbes.com/sites/caroltice/2014/08/14/7-fastfood-restaurantchains-that-rake-in-2m-per-store/#5ff411a9ff82

# Appendices

## Appendix A

CDC contributing factors' definitions and their relationship to one or more of the top five risk factors that lead to foodborne illness and outbreaks in foodservice; the top five risk factors are (1) **food from unsafe source,** (2) **poor personal hygiene,** (3) **inadequate cooking,** (4) **improper holding/time and temperature**, and (5) **contaminated equipment/protection from contamination.**

### Contamination Factors

Factors that introduce or otherwise permit contamination; contamination factors relate to how the etiologic agent got onto or into the food.

**Risk Factor Number 1—Toxic Substance Part of the Tissue (e.g., Ciguatera)** A natural toxin found in a plant or animal or in some parts of a plant, animal, or fungus OR a chemical agent of biologic origin that occurs naturally in the vehicle or bioaccumulates in the vehicle prior to or soon after harvest.

Examples of this type of contributing factor include ciguatera fish poisoning due to consumption of marine fin fish or mushroom poisoning due to consumption of toxic mushrooms.

**Risk Factor Number 1—Poisonous Substance Intentionally/Deliberately Added (e.g., Cyanide or Phenolphthalein Added to Cause Illness)** A poisonous substance intentionally or deliberately added to a food in quantities sufficient to cause serious illness. Poisons added because of sabotage, mischievous acts, and attempts to cause panic or to blackmail a company fall into this category.

© Springer Nature Switzerland AG 2020
H. King, *Food Safety Management Systems*, Food Microbiology and Food Safety, https://doi.org/10.1007/978-3-030-44735-9

This contributing factor only applies to poisonous substances, not to physical substances added to food.

**Risk Factor Number 5—Poisonous Substance Accidentally/Inadvertently Added (e.g., Sanitizer or Cleaning Compound)** A poisonous substance or chemical agent accidentally/inadvertently added to the vehicle. This addition typically occurs at the time of preparation or packaging of the vehicle.

Examples of this type of contributing factor include sanitizer or cleaning compound added to food or chemicals that reach foods from spillage or indiscriminate spraying. Misreading labels, resulting in either mistaking poisonous substances for foods or incorporating them into food mixtures, also falls into this category.

**Risk Factor Number 5—Addition of Excessive Quantities of Ingredients that Are Toxic in Large Amounts (e.g., Niacin Poisoning in Bread)** An approved ingredient in a food but accidentally added in excessive quantities so as to make the food unacceptable for consumption.

Examples of this type of contributing factor include excessive amounts of nitrites in cured meat or excessive amounts of ginger powder in gingersnaps.

**Risk Factor Number 5—Toxic Container (e.g., Galvanized Containers with Acidic Foods)** Container or pipe holding or conveying the implicated food is made of toxic substances. The toxic substance either migrates into the food or leaches into solution by contact with highly acidic foods.

One example of this type of contributing factor is a toxic metal (e.g., zinc-coated) container used to store highly acidic foods. For this contributing factor, there may be confusion between foodborne outbreaks and waterborne outbreaks. If the outbreak is waterborne, the contributing factors should be listed in the waterborne section, not in this foodborne section. In general, waterborne disease included contamination occurring in the source water or in the treatment or distribution of water to the end consumer. For example:

- If water enters a contaminated drink mix/soda machine or if there is a problem with the internal plumbing of the machine resulting in contamination (e.g., cross-connections, backflow of carbonated water resulting in copper leaching)—it's waterborne and should not be entered in the foodborne section.
- If ice is made with contaminated water—it's waterborne and should not be entered in the foodborne section.
- If ice is already made and then it becomes contaminated because it was stored in a toxic container—it is a foodborne outbreak and it would be appropriate to list C5 as a contributing factor.

**Risk Factor Number 3—Contaminated Raw Product (Food Was Intended to Be Consumed After a Kill Step)** The vehicle or a component of the vehicle contained the agent when it arrived at the point of final preparation or service. This contributing factor applies to foods intended to be consumed after undergoing a kill step (such as cooking to the required temperature), but the food processing step was insufficient to lower the levels of the pathogen below an infectious dose.

Examples of this type of contributing factor include a hamburger that was ordered well-done or medium-well but subsequently undercooked when it arrived at final preparation or raw chicken that was contaminated with *Salmonella*, which was then unintentionally undercooked.

**Risk Factor Number 1—Contaminated Raw Product (Food Was Intended to Be Consumed Raw or Undercooked/Under-Processed (e.g., Raw Shellfish, Produce, Eggs))** Contaminated products are ingested raw without being first subjected to a cooking step or another form of a kill step sufficient to kill any pathogens present. This contributing factor applies to foods intended to be consumed raw, as well as foods intended to be consumed after mild heating or another process that does not ensure pathogen destruction. Mild heating means heated to time/temperature exposures insufficient to kill vegetative forms of pathogenic bacteria or denatured proteins.

Examples of this type of contributing factor include mildly heated hollandaise sauce containing raw egg yolk, a hamburger or steak ordered to be prepared rare, raw milk, raw oysters or other shellfish, raw produce, or unpasteurized cider or juices.

**Risk Factor Number 1—Foods Originating from Sources Shown to Be Contaminated or Polluted (Such as a Growing Field or Harvest Area) (e.g., Shellfish)** Foods obtained from sources shown to be contaminated, such as shellfish from sewage-polluted waters, crops watered by contaminated irrigation water, or produce grown in contaminated soil.

Note: Formal traceback may support or confirm the identification of this contributing factor. This factor would typically be cited along with another contamination factor.

**Risk Factor Number 5—Cross-Contamination of Ingredients (Does not Include Ill Food Workers)** Pathogen transferred to the ingredient by contact with contaminated worker's hands, equipment, or utensils or by drippage or spillage. If worker's hands were the mode of contamination, the worker was not infected with or a carrier of the pathogen.

Examples of this type of contributing factor include:

- Contaminated raw poultry was prepared on a cutting board; later, a ready-to-eat food was cross-contaminated because it was prepared on this same cutting board without intervening cleaning.
- A worker's hands became contaminated by raw foods; subsequently, a ready-to-eat food was cross-contaminated because the worker's hands touched this ready-to-eat food without intervening hand washing.

- Cloths, sponges, and other cleaning aids were used to clean equipment that processed contaminated raw foods. Before their next use, these cleaning items were not disinfected; instead, these cleaning items are used to wipe surfaces that come in contact with foods that are not subsequently heated.
- Contaminated raw foods touch or fluids from them drip onto foods that are not subsequently cooked.

This contributing factor only applies to foods that are cross-contaminated by other ingredients.

**Risk Factor Number 2—Bare-Hand Contact by a Food Handler/Worker/ Preparer Who Is Suspected to Be Infectious (e.g., with Ready-To-Eat-Food)** A food worker suspected to be infectious uses his or her bare hands to touch or prepare foods that are not subsequently cooked. The term "infectious" is an all-inclusive term used to describe all persons who are colonized by, infected with, a carrier of, or ill due to a pathogen. This is a typical situation that precedes outbreaks caused by Norovirus or staphylococcal enterotoxins.

**Risk Factor Number 2—Glove-Hand Contact by a Food Handler/Worker/ Preparer Who Is Suspected to be Infectious (e.g., with Ready-To-Eat-Food)** A food worker suspected to be infectious uses his or her gloved hands to touch or prepare foods that are not subsequently cooked. The term "infectious" is an all-inclusive term used to describe all persons who are colonized by, infected with, a carrier of, or ill due to a pathogen. This is a typical situation that precedes outbreaks caused by Norovirus or staphylococcal enterotoxins.

**Risk Factor Number 2—Other Modes of Contamination (Excluding Cross-Contamination) by a Food Handler/Worker/Preparer Who Is Suspected to be Infectious** A food worker suspected to be infectious contaminates the food by another mode of contamination other than bare-hand contact or glove-hand contact, or an epidemiological/environmental investigation determines that an infectious food worker contaminates food with his or her hands, but the investigation is unable to determine whether or not the food worker was wearing gloves during food preparation. This contaminated food is subsequently not cooked.

Examples of this contributing factor include:

- Epidemiological or environmental investigation determines that an infectious food worker contaminates food with his/her hands but is unable to determine whether or not actual bare-hand contact or glove-hand contact contaminated the food.
- In Norovirus outbreaks, an ill food worker's aerosolized vomitus contaminates ready-to-eat food.

**Risk Factor Number 2—Foods Contaminated by Non-food Handler/Worker/ Preparer Who Is Suspected to be Infectious** A person other than a food handler/ worker/preparer suspected to be infectious contaminates ready-to-eat foods that are later consumed by other persons, resulting in spread of the illness. A non-food handler/worker/preparer is any person not directly involved in the handling or preparation of food before service. This is a typical situation when an ill person attends an event and contaminates ready-to eat-foods in a buffet line by handling food before someone else consumes it. The original ill person is identified as a source of the pathogen.

One example of this type of contributing factor is a when healthy food worker prepares pizza, which arrives pathogen-free. A mother (a non-food worker) rearranges pizza slices onto plates before serving the slices to a group of children at a birthday party (regardless of setting—it could be at a home or a restaurant). These children subsequently develop foodborne illness and the mother is identified as a source of the pathogen.

**Risk Factor Number 5—Storage in Contaminated Environment (e.g., Storeroom, Refrigerator)** Storage in a contaminated environment (such as a storeroom or refrigerator) leads to contamination of the food vehicle or an ingredient in the vehicle. This contributing factor only applies to stored foods that were contaminated directly by environmental sources, not contamination by other foods. This usually involves storage of dry foods in an environment where contamination is likely from overhead drippage, flooding, airborne contamination, access of insects or rodents, and other situations conducive to contamination.

This contributing factor only applies to food contaminated during storage, not foods contaminated during preparation or service.

**Risk Factor Number 5—Other Sources of Contamination** A form of contamination that does not fit into the above categories. Physical substances added intentionally or deliberately also fall into this category. Objects can get into food either from lack of removal of seeds or other hard particles or from objects in the soil.

Examples of this contributing factor include glass shards intentionally or deliberately added to food, food in an uncovered bowl contaminated by flies, or food being washed or soaked in a food preparation sink that gets contaminated by sewage backflow from the sink's pipes.

**Proliferation/Amplification Factors (*Bacterial Outbreaks Only*)**

Factors that allow proliferation or amplification of the etiologic agents; proliferation/amplification factors relate to how bacterial agents were able to increase in numbers and/or produce toxic products before the food was ingested.

**Risk Factor Number 4—Food Preparation Practices that Support Proliferation of Pathogens (During Food Preparation)** During food preparation, one or more improper procedures occurred (such as improper or inadequate thawing) that allowed pathogenic bacteria and/or molds to multiply and generate to populations sufficient to cause illness or to elaborate toxins if toxigenic.

Examples of this type of contributing factor include:

- Improper thawing (such as allowing frozen food to thaw at room temperature or leaving frozen foods in standing water for prolonged periods) allows pathogens on the surface of the food to multiply and generate.
- Prolonged preparation time (such as prolonging preparation time by preparing too many foods at the same time) allows pathogens to multiply and generate.

**Risk Factor Number 4—No Attempt Was Made to Control the Temperature of Implicated Food or the Length of Time Food Was Out of Temperature Control (During Food Service or Display of Food)** During food service or display of food, no attempt was made to control the temperature of the implicated food or no attempt was made to regulate the length of time food was out of temperature control.

Examples of this type of contributing factor include leaving foods out at ambient temperature for a prolonged time at a church supper or no time and temperature control on a buffet line.

**Risk Factor Number 4—Improper Adherence of Approved Plan to Use Time as a Public Health Control** Food out of temperature control for more than the time allowed under an agreed-upon and preapproved plan by a regulatory agency to use time as a public health control.

Examples of this type of contributing factor include:

- Foods are placed on a buffet table that is not capable of maintaining proper hot or cold temperatures. The establishment has a plan approved by a regulatory agency to use time as a public health control. The plan allows foods to be displayed for service on the buffet line at ambient temperature and then discarded after 4 h. However, the food is held on the buffet table for longer than 4 h (either inadvertently or intentionally).
- A facility negotiates a plan with a regulatory agency to use time as a public health control. The facility improperly adheres to the plan because some of the dishes that the facility serves are traditionally held and served at room temperature longer than the time allowed in the approved plan.

**Risk Factor Number 4—Improper Cold Holding Due to Malfunctioning Refrigeration Equipment** Malfunctioning refrigeration equipment (such as improperly maintained or adjusted refrigerators) causes foods to be held at an improper cold holding temperature or walk-in cooler malfunction causes elevated temperatures of food.

Examples of this type of contributing factor include:

- The reach-in (or walk-in) refrigerator unit temperature is not monitored and stays consistently higher than 41°F, causing elevated temperatures of food.
- A broken or torn door gasket causes air leakage in a reach-in refrigerator and subsequently food remains above 41°F.

**Risk Factor Number 4—Improper Cold Holding Due to an Improper Procedure or Protocol** Improper cold holding temperature because of an improper procedure or protocol (such as an overloaded refrigerator or inadequately iced salad bar).

Examples of this type of contributing factor include potentially hazardous foods such as tuna salad or egg salad stacked above the top levels of the cold holding wells in a deli sandwich cold holding unit.

**Risk Factor Number 4—Improper Hot Holding Due to Malfunctioning Equipment** Equipment meant to be used for hot holding malfunctions and causes foods to be held at an improper hot holding temperature.

Examples of this type of contributing factor include a steam table that is improperly maintained or adjusted and causes food to be held at improper hot holding temperatures.

**Risk Factor Number 4—Improper Hot Holding Due to Improper Procedure or Protocol** Improper hot holding temperature because of an improper procedure or protocol.

Examples of this type of contributing factor include:

- An inadequate number of Sterno cans are used for holding foods hot in chafing dishes.
- Exhausted Sterno cans are not replaced under chafing dishes that hold hot foods.
- Steam table was not turned on.

**Risk Factor Number 4—Improper/Slow Cooling** Foods refrigerated in large quantities or stored in devices where temperature is poorly controlled allow pathogens to multiply. Improperly cooling foods are those outside of these parameters: cooling foods from 135°F to 70°F within 2 h and from 70°F to 41°F within the next 4 h.

Examples of this type of contributing factor include:

- Foods are refrigerated in large quantities (i.e., in large masses or as large volumes of foods in containers) that do not allow proper cooling.
- Foods are stored in containers with tight-fitting lids, leading to inadequate air circulation and thus improper cooling.

**Risk Factor Number 4—Prolonged Cold Storage**  This situation is a concern for psychrotrophic pathogenic bacteria (e.g., *Listeria monocytogenes, Clostridium botulinum* type E, *Yersinia enterocolitica, Aeromonas hydrophila*) that multiply over sufficient time at ordinary refrigerator temperatures and generate to populations sufficient to cause illness or elaborate toxins if toxigenic (e.g., *C. botulinum*).

Examples of this type of contributing factor include:

- Holding foods prepared in a foodservice establishment in cold storage for more than 7 days
- Holding open containers of commercially prepared foods for several weeks

**Risk Factors Number 1 and 4—Inadequate Modified Atmosphere Packaging (e.g., Vacuum-Packed Fish, Salad in Gas-Flushed Bag)**  Food stored in a container that provided an anaerobic environment. These factors create conditions conducive to growth of anaerobic or facultative bacteria in foods held in hermetically sealed cans or in packages in which vacuums have been pulled or gases added.

All anaerobic bacteria must have a low oxygen reduction potential to initiate growth, but this factor is restricted only to foods that are put into the sealed package or container. Some restaurants seek HACCP plan approved to perform Modified Atmosphere Packaging (MAP).

**Risk Factor Number 1—Inadequate Processing (e.g., Acidification, Water Activity, Fermentation)**  Inadequate non-temperature-dependent processes (such as acidification, water activity, fermentation) that do not prevent proliferation of pathogens, which multiply and generate populations sufficient to cause illness.

Examples of this type of contributing factor include:

- Insufficient acidification (low concentration of acidic ingredients) in home-canned foods
- Insufficiently low water activity (low concentration of salt) in smoked/salted fish
- Inadequate fermentation (starter culture failure or improper fermentation conditions) in processed meat or processed cheese

**Risk Factor Number 4—Other Situations that Promote or Allow Microbial Growth or Toxin Production (Please Describe)**  A factor that promotes growth, proliferation, amplification, or concentration of etiologic agents but does not fit into any of the other defined categories; the factor should be specified.

One example of this type of contributing factor is a box of tomatoes that was unknowingly contaminated by *Salmonella* before its arrival at a restaurant. Soon after the delivery, some of the tomatoes were served to customers but these customers did not become ill. Some of the other tomatoes from the box were not served soon after delivery—instead, these tomatoes were allowed to ripen at room temperature for several days, which allowed the *Salmonella* to amplify. Customers who ate the room-ripened tomatoes became ill. Although allowing intact tomatoes to ripen at room temperature is not a Food Code violation, this process likely led to bacterial proliferation.

## Survival Factors (*Primarily Bacterial Outbreaks*)

Factors that allow survival or fail to destroy or inactivate the contaminant; survival factors refers to processes or steps that should have eliminated or reduced the microbial agent but did not.

**Risk Factor Number 3—Insufficient Time and/or Temperature During Cooking/Heat Processing (e.g., Roasted Meats/Poultry, Canned Foods, Pasteurization)** Time/temperature exposure during initial heat processing or cooking inadequate to kill the pathogen under investigation. In reference to cooking, it refers to the destruction of vegetative forms of bacteria, viruses, and parasites, but not bacterial spores and sometimes not bacterial toxins (e.g., heat-resistant ones). If the food under investigation was retorted, then spore-forming bacteria would be included.

This does not include inactivation of preformed heat-stable toxins or destruction of bacterial spores during cooking.

**Risk Factor Number 3—Insufficient Time and/or Temperature During Reheating (e.g., Sauces, Roasts)** Time/temperature exposure during reheating or heat processing of a previously cooked or heated food (which has often been cooled overnight) inadequate to kill the pathogens.

This does not include inactivation of preformed heat-stable toxins.

**Risk Factor Number 4—Insufficient Time and/or Temperature Control During Freezing** Insufficient time and/or temperature control during freezing of foods such as fish, which may be frozen before raw service.

One example of this type of contributing factor is when there is insufficient time and/or temperature control during freezing: Pacific red snapper is the implicated food in an outbreak of Anisakis infection. The snapper was not frozen before service in raw sushi or the investigation revealed that the time and temperature required to kill parasites (-31°F for 15 h or 4°F for 7 days) was not used.
Freezing is currently used for parasite destruction in fish served raw. In the future if it is determined that freezing can be used for pathogen destruction in other situations, this factor would be cited if established procedures are not implemented or are implemented incorrectly. Some species of tuna are not susceptible to harboring parasites of concern, so freezing is not necessary.

Care should be taken in determining if freezing would have been an appropriate pathogen destruction process for the fish in question before citing S3.

**Risk Factor Number 1—Insufficient or Improper Use of Chemical Processes Designed for Pathogen Destruction** Insufficient or improperly used chemical processes (such as acidification, salting, and cold smoking) allow pathogens to survive.

Examples of this type of contributing factor include:

- Inadequate acidification (such as insufficient quantity or concentration of acid) of canned tomatoes results in pathogen survival.
- Inadequate cold smoking of meat (such as insufficient time of contact of the smoke with the meat) results in pathogen survival.

**Risk Factor Number 1—Other Process Failures that Permit the Agent to Survive (Please Describe)** Other forms of survival. A form of survival that does not fit into the above categories; the factor should be specified. Failures of other processes (such as subjecting foods to irradiation, high pressure, drying conditions) that then permit pathogens to survive. Specify the survival factor.

# Appendix B

**A Regulator's Perspective on Food Safety Management Systems to Achieve Active Managerial Control in Foodservice Operations: Shannon Mckeon, REHS Environmental Health Specialist III**

When I first started inspecting restaurants 14 years ago, I could get a sense of how an inspection was going to go by the reaction I got from restaurant staff when I introduced myself as a health inspector. A smile and a warm welcome was usually a good sign that the operator was comfortable with an evaluation of their restaurant and understood that we had a shared interest in food safety. A grimace, a panicked look, and—yes—even tears gave me the impression that the owner was anticipating negative feedback. This was, at times, an early indicator that I was about to have a challenging inspection. Luckily, I'm greeted with a lot more warm welcomes now than I was 14 years ago. I believe that this can be attributed to two things: (1) the relationship between the restaurant operator and the health inspector has shifted from one of citation and enforcement to more of a partnership, with a common goal of providing safe food to the public, and (2) restaurant owners have confidence in their operation because they have incorporated Active Managerial Control (AMC) into their food safety practices. In my observation, this shift has resulted in a more cooperative assessment, improvement of foodservice operations, and a reduction in the occurrence of the risk factors for foodborne illness (as evidenced by risk factor studies conducted in my jurisdiction in 2005, 2010, and 2016). This Appendix will serve to discuss the value of AMC to both the foodservice operation and the regulatory authority, explore the regulator's responsibility to assess and promote AMC in foodservice establishments, provide examples of collaborating with operators to encourage and offer incentives for practicing AMC, and share how to create a culture of AMC promotion among regulatory staff.

## The Value of Active Managerial Control

The common goal between the regulatory authority and foodservice operators is to ensure customer safety. One might assume that a regulator's sole concern is compliance and that an operator's bottom line is profit. However, in reality, both want customers to have access to safe, quality food. Further, both parties understand that this standard can be attained while also bringing value to the business needs of the operation. There are several business benefits to implementing AMC, as outlined in Annex 4 of the FDA Food Code and throughout this book. In addition to conforming to regulations, foodservice establishments that incorporate AMC into their operation benefit through:

- Increased employee awareness and participation in food safety
- Consistent product preparation
- Increased quality
- Reduced product loss
- Improved inventory control
- Increased profit

As a regulator promoting AMC to foodservice operators, it is important to share these business-level benefits to support buy-in to the value of AMC. AMC requires dedication of time, energy, and resources; however, the long-term benefits that come with strengthening Food Safety Management Systems (FSMS) make AMC well worth the investment.

I've had the opportunity to see AMC concurrently benefit both a foodservice operation and the regulatory authority. I came to believe in the value of AMC from a regulatory perspective many years ago when I was assigned a facility that chronically struggled to maintain proper cold holding temperatures for Time/Temperature Control for Safety (TCS) foods. Each of my visits would inevitably result in a citation. After several inspections demonstrated non-compliance, enforcement action was initiated. However, progressive enforcement proved ineffective in attaining compliance. Instead of escalating enforcement activities, it was determined that employee training would be the best course to help facilitate and encourage compliance. The training was presented in a way that helped the operators and their staff understand how to achieve AMC over cold holding. After this training, the business demonstrated proper cold holding temperatures at their follow-up inspection and maintained proper cold holding at subsequent routine inspections.

The important discussion piece from this story is why and how did this facility finally achieve compliance? What was different in this scenario that promoted long-term controls to ensure compliance? A few factors played into the success of this approach. First, and perhaps most importantly, FSMS related to the control of the risk factor was finally implemented. A potential shortcoming of health inspections and their corresponding reports is that **they elicit a reactive—rather than proactive—response**. Inspectors identify and document a violation and seek immediate correction of any risk factor identified. However, in order to achieve long-term control over a risk factor, it is important for foodservice operators to understand how to proactively monitor and control hazards through the practice of AMC rather than relying on an inspector's periodic feedback. In this example, rather than employing continued citation or enforcement focused on immediate and short-term correction, AMC was presented to the foodservice staff to increase their understanding of the violation and to empower them to ensure long-term compliance. AMC was also presented in a context that was very specific to this establishment. Discussions of the concepts of AMC with foodservice operators are valuable; however, that value is multiplied when AMC is presented in a way that is relevant and useful to a specific foodservice business.

Also instrumental in the success of this approach was the inclusion of all employees at the training event. Oftentimes during inspections and enforcement, violations and control measures are only discussed with the manager or person in charge at the time of the inspection. This limits the understanding to one person in the operation and does not involve the front-line staff who may be able to make the greatest impact on food safety by changing their practices. AMC cannot be effectively achieved by one person in an operation: it requires the integration of all staff into FSMS. Engaging the staff at this training gave them an increased understanding of the risk factor and the importance of its control. This, in turn, helped support buy-in for staff involvement or process change. Additionally, the description of AMC specific to this establishment and the identified risk factor helped demonstrate the role of staff in long-term compliance. This training prepared staff for future assignment of specific tasks related to FSMS and direction for intervention strategies such as monitoring of controls and corrective actions. Furthermore, a presentation from the health department carries a certain impact with the staff. The fact that the health department recognizes the importance of addressing a specific issue at their facility lends a sense of urgency and importance to the issue. Having AMC heavily suggested in this case—or even apparently mandated—by the health department helps support the manager's initiatives to incorporate AMC practices into staff roles and responsibilities.

Lastly, although this training was technically a component of enforcement proceedings, the use of training in place of traditional enforcement activities facilitated a more collaborative—rather than punitive—relationship, enabling the health department to be viewed as partnering with the foodservice business to impart knowledge, provide valuable resources, and support the staff in their efforts to incorporate AMC into their every day practices. Developing a more comfortable and trusting relationship with the health department encourages the foodservice business to utilize their inspector as a resource in the future to help identify or correct issues that may arise or are anticipated. This is an example of the shift from enforcer to partner. In my experience, this shift has resulted in improved collaboration between industry and regulators and an increased commitment to regulatory compliance and attention to food safety. The example above helped me realize the value of AMC not only for a foodservice operation but also for the regulatory authority. AMC may serve also to benefit the health department through:

- Achievement of long-term compliance of regulated foodservice businesses
- Reduction of time spent at inspections due to increased compliance
- Reduction in the need for enforcement action
- Reduction in inspection frequency based on risk, enabling the regulatory authority to shift its focus to riskier establishments
- Potential reduction in complaints linked to foodborne illness and their risk factors

The story I've related above—combined with an appreciation for the overall value of AMC—was the catalyst that would lead to the development of AMC promotion and recognition programs in my jurisdiction.

### The Regulator's Responsibility to Assess and Promote Active Managerial Control

As discussed throughout this book, in order to make a significant impact on food-borne illness risk factors, foodservice businesses should incorporate AMC's purposeful, preventative food safety management practices via FSMS into their operation. But where does a regulator fit into this? Annex 4 of the FDA Food Code outlines a twofold role for regulators in supporting and encouraging AMC:

1. To incorporate assessment of the degree of AMC a foodservice business demonstrates into the inspection process
2. To assist foodservice businesses in developing and implementing voluntary strategies to prevent the occurrence of risk factor violations

The regulator's role in facilitating AMC in an operation should not be passive but, rather, it should be active in assessment and promotion. Having an inspection process that is designed to assess the degree of AMC demonstrated by a foodservice business is more challenging than it may seem. Compliance documented on a health department inspection report alone is not a direct indicator of proactive FSMS; therefore, it does not have a direct correlation to AMC. Because of this, regulators must incorporate additional observations and questioning into their inspection process in order to identify and evaluate the AMC of an establishment. But first, a regulator must have a deep understanding of the concept of AMC and the ability to recognize AMC practices in the field in order to conduct an effective assessment. When my jurisdiction initially committed to dedicating effort and resources to the promotion of AMC, it became clear that our regulatory officials would benefit from a consistent understanding of the concept of AMC and how to implement AMC assessment techniques. So we first trained regulatory staff on AMC to provide them with the knowledge, skills, and ability to identify, assess, and encourage AMC in foodservice operations. The training program for new regulators has been updated to include the theory, identification, and promotion of AMC.

Once regulators were empowered to assess and provide feedback regarding the level of AMC in foodservice establishments, it was important to ensure that these assessments were as simple as possible and ingrained in the inspection process. Because inspectors are constantly faced with increasing workloads and diminishing time and resources, it is important to consider these limitations when designing assessment protocols. Defining methods for efficiently and effectively assessing AMC that will not significantly lengthen inspection time will facilitate increased regulator participation. Providing resources that aid in AMC assessment will also save time and assist in standardizing assessments. Such resources include guidance documents (e.g., a list of common AMC practices associated with the risk factors) or feedback tools (e.g., an "AMC Report Card" to provide quick and organized feedback).

With an inspection process designed to assess the degree of AMC demonstrated by an operation, a regulator may then use the information gleaned to assist operators in developing and implementing voluntary strategies to prevent the occurrence of risk factor violations. A first step (similar to that used to train regulatory staff) is introducing the concepts of FSMS and AMC to the foodservice operators. In my experience, even some operators who are actively practicing AMC are not familiar with the term. Thus, it is important to introduce the concept and help them to identify and promote AMC practices within their facilities. Familiarizing operators with AMC can be achieved both with the jurisdiction's foodservice community at large and at the level of the individual foodservice establishment. For example, a newsletter is a great way to communicate with the entire regulated community. When published regularly, a newsletter is a valuable tool for sharing food safety information and proactive FSMS with foodservice operators in a low-pressure, collaborative environment. Additionally, information and resources related to AMC may be placed on the website so that it is easily accessible for all operators.

To provide AMC information to individual operations, regulators may want to consider ways to incorporate AMC introductions into the inspection process and documentation. Relating observations and suggestions for improvement made during the inspection to AMC will make AMC a part of the inspection vernacular and hopefully build interest in developing, continuing, and/or strengthening AMC practices. Further, continued adherence to good food safety practices is of equal importance to the mitigation of risky behaviors. For this reason, it helps to recognize good practices during the inspection, praising and encouraging the continuation of these practices in both conversation and the inspection report comments. This serves to inform the operator of practices that are contributing to the reduction in risk factor violations and lend a health department endorsement to these behaviors.

A foodservice operator who understands the concepts of AMC is now ready to receive feedback specific to their operation. If AMC practice at an establishment has room for improvement, those suggestions may be provided at this time—cushioned, whenever possible, with the positive feedback. Consider, for example, a hypothetical facility that utilizes a log to document food temperatures. The inspector notices that the log does not have space to document corrective actions taken in the case that an improper temperature is noted. The regulator would recognize and praise the value of the log while also noting the opportunity to strengthen the facility's FSMS and documentation by adding space in the log for marking corrective actions.

It is important to remember when making recommendations for strengthening systems to clearly differentiate between recommendation and requirement. As the regulatory authority, it is likely that most of what we communicate to an operator may come across as a mandate. Although highly recommended, most AMC practices are voluntary and should be presented as such to avoid confusion between regulatory requirements and suggested best practices. It is also helpful from a regulatory perspective to make note of AMC practices demonstrated by a foodservice business and remember to inquire about progress with AMC at each subsequent inspection. This keeps conversations regarding AMC active and may serve to inspire continuous and/or improved AMC practices if an updated request is anticipated.

Having the appropriate resources necessary to facilitate AMC will also be an integral part of encouraging a foodservice business to adopt these practices. For example, an AMC self-assessment tool like one designed in an FSMS described in this book is a helpful resource to provide to all facilities. This tool enables an establishment to conduct a baseline assessment of the degree of AMC in their operation and help focus improvement efforts where needed. Many large chains have the training materials, standard operating procedures (SOPs), and logs necessary to institute FSMS; however, the majority of small-chain or single-owner foodservice establishments do not have access to these types of resources. The regulator may encourage these facilities to develop their own tools and materials; however, I encourage developing materials and templates within the health department.

Regulators have a vast knowledge of the components of an effective food safety management system as well as an understanding of how to effectively communicate with the regulated community. Thus, regulators can bring that expertise to bear in developing tools and training materials that are accurate, focused, and easy to navigate. Templates may be used as-is; alternatively, the regulators may use their historical knowledge of the foodservice business to assist the operators in tailoring the tools to meet their establishment's specific needs, based on their food preparation processes, equipment, staffing, etc. Additionally, because AMC requires a team effort, it will be important to engage and train staff on AMC principles and practices. Training tools such as fact sheets, signs, posters, and presentations may be provided to operators to help facilitate staff training. A comprehensive collection of the most important AMC tools and training materials may be provided in the form of an AMC Toolkit or Starter Kit to encourage operators to begin to utilize AMC practices without overwhelming them with an abundance of materials.

If the health department has sufficient resources to do so, offering formal AMC training to foodservice operators and staff would also be beneficial. Through their inspection experience and extensive food safety training, regulators have been exposed to countless examples of effective AMC and are uniquely positioned to present this information to foodservice staff. Having a training offsite and at a scheduled time enables the foodservice business operator to focus on learning AMC practices more readily than would be the case during an inspection. This may also serve to better the partnership between foodservice businesses and regulators, who may be seen as teachers sharing their knowledge to assist operators in improving the safety and success of their operation.

When undertaking AMC promotion efforts, regulators should devote some fore-thought to data collection with the hope of being able to measure the prevalence of AMC in their jurisdiction so that they can demonstrate an improvement associated with their efforts. Metrics are very important to demonstrate public health out-comes. Standard health inspection data alone cannot be utilized to determine the prevalence of AMC in a jurisdiction: although it shows compliance, it does not necessarily demonstrate that such compliance was correlated to instituting proac-tive FSMS.

There are several ways to collect data regarding the prevalence of AMC, each with its own strengths and limitations. The easiest and most accurate way to capture whether a facility demonstrates AMC would be to incorporate an assessment item into the inspection process and report form. This would ensure that AMC assess-ment occurs during routine inspections, would not require the completion of addi-tional forms or data entry into a separate database, and enables assessment by a trained regulator. A potential obstacle to this approach is that it is not always easy or feasible to add an inspection item to a form or adjust within inspection software, which may preclude some jurisdictions from adopting this approach. A secondary data collection form and database may be created to collect and analyze the occur-rence of AMC in a jurisdiction; however, this requires completion of a separate form and additional data entry. The separate form and data entry translate to additional time and resources. Lastly, regulators may survey their foodservice operations, ask-ing them to self-report their AMC practices; however, this self-evaluation may not have the accuracy of a regulator's assessment.

Whatever method is chosen, having a baseline assessment of AMC in a jurisdic-tion will help define the current state of AMC before promotion efforts. A regula-tor's jurisdiction may then institute AMC promotion and training activities and track the impact of their efforts through the analysis of future re-assessments. This will also assist regulators' teams in applying targeted interventions and training if a discrepancy is identified in AMC associated with certain facility types, food pro-cesses, risk factors, or other assessment categories.

## AMC Recognition Programs Can Improve Food Safety in Foodservice Businesses

The first program in my jurisdiction that was developed to promote and incentivize AMC stemmed from the desire to recognize and highlight those foodservice employees (not the business itself) who went above and beyond normal practices to enhance food safety in their facility. One day, an inspector in my jurisdiction noted a foodservice employee proudly wearing a lapel pin on their uniform and inquired about the meaning. It was a customer service pin, but it sparked the idea that attention to food safety—or, more specifically, the demonstration of AMC—should be awarded in a way that is public-facing and demonstrates endorsement of AMC practices. From there, our jurisdiction created the AMC award, which may be awarded at the time of an inspection to a person in charge who effectively demonstrates AMC. We engaged our staff to determine program criteria and to define what exactly they believed qualified as the effective demonstration of AMC. Our staff determined that a person in charge may be presented with an AMC award if they demonstrate knowledge regarding safe food handling practices, have an inspection absent of critical violations, and utilize FSMS to ensure food safety. To make this award impactful, the health department assigned benefits to the program. An award recipient:

- Receives an accolade lapel pin which may be worn to showcase their commitment to food safety and the health department recognition thereof
- Is featured in the quarterly newsletter, *Food for Thought*, which is distributed to all foodservice establishments in our jurisdiction (3,700 and growing)
- Is highlighted on the AMC page of the health department website
- Receives a summary of their AMC practices and achievements in their inspection report, which is available to the public and facility or corporate management

When we initiated this program, we knew immediately that we were on to something special. Foodservice managers were thrilled to be recognized for their hard work and expressed how much the recognition meant to them. Some operators were moved to tears, stating that this was the first time in their career that they had been recognized for their efforts. The award additionally served to strengthen the relationship between the inspector and the operator. Those who were awarded were excited to show off their AMC practices at future inspections and were more comfortable with the inspection process as a whole. Regulators have even experienced employees vying over who would have the opportunity to accompany them through the inspection, because all the employees in the facility wanted their chance to receive the award. This award helped to re-frame the tone of an inspection from a visit that was feared to a visit that was excitedly anticipated as a way to show off operators' hard work and achievements. This award was a great starting point for recognizing and incentivizing AMC, but it missed the mark on one big element: the fact that it takes a team—not a single person—to fully achieve AMC. So, while maintaining our program to recognize individuals for their contributions to AMC, we developed another, more robust program to recognize an entire operation and all of the associated staff for exceptional FSMS called **STAMP**, or **Safety Through Actively Managing Practices.**

The current STAMP program strives to accomplish a more holistic analysis of AMC in a foodservice operation, ensuring that FSMS are practiced by the whole team in a foodservice business and incorporated into all aspects of the operation. Whereas the AMC award may be issued at the time of an inspection during which a manager demonstrates food safety knowledge, has no critical violations, and practices some degree of AMC, a prospective STAMP enrollee must go through a rigorous application and evaluation process in order to establish that AMC is fully incorporated into their operation. Within their application, an operator outlines formal food safety policies and procedures, staff training, monitoring activities, and third-party or self-inspections conducted.

Additionally, an AMC self-assessment is conducted by the foodservice operator, which scores an operation based on AMC best practices and requires a passing score to proceed to the next step in the enrollment process, an On-Site Assessment (OA). An OA is an on-site evaluation conducted by an Environmental Health Specialist to verify that all listed policies and procedures, staff training, and monitoring activities exist and are actively and effectively used. A review of previous inspections must demonstrate exceptional performance in the reduction of foodborne illness risk factors. If an operation successfully passes a review of their application and their On-Site Assessment, they are approved as an enrollee in the STAMP program.

Considering the difficulty of participation in this program, it was important for us to provide benefits to the establishment that would appropriately incentivize enrollment. It is the hope of the Department that these incentives will encourage foodservice businesses that demonstrate AMC to continue safe food handling practices and inspire operations lacking FSMS to explore their integration. A facility that successfully enrolls in STAMP is one that prioritizes food safety and invests time, money, resources, and attention to protecting their customers by controlling foodborne illness risk factors. We considered it important to make consumers aware of establishments that invest in their protection through food safety. For that reason, we decided to make enrollment in this program highly visible to the public. A STAMP program enrollee receives:

- A STAMP decal (Fig. 1) to be placed on the window or door of their operation
- A certificate of enrollment that may be displayed within their establishment
- A letter of commendation from the Director of Environmental Health, celebrating their dedication to food safety and demonstration of exemplary best practices
- Recognition in the quarterly newsletter, *Food for Thought*
- Inclusion on the health department website in a list and searchable map of STAMP participants
- A reduction in risk categorization and therefore inspection frequency, based on the increased protection of the public due to their demonstration of AMC

I want to highlight two of the benefits: one is most important to the foodservice business, and the other is most important to the regulatory authority. The STAMP decal may be the benefit that brings the most value to the operator. The decal is visible from the outside of the facility and facilitates customer awareness of the establishment's participation in the program. In lieu of issuing letter grades, scores, or other indicators of performance, our jurisdiction makes all inspection reports available online. Although this allows customers to view our inspection reports in their entirety, it does not enable passersby to make an at-a-glance determination of the establishment's performance. The STAMP decal is the visible indicator of food safety performance in the county; it demonstrates that the facility performs well

**Fig. 1** The Food Safe STAMP decal to demonstrate food safety for operations whose AMC practices have been evaluated and approved by the health department

during its inspections and that their AMC practices have been evaluated and approved by the health department. Customers may dine in confidence that AMC practices are in place to ensure food safety. We anticipate that consumers will actively seek out facilities that enroll in this program. This will encourage other foodservice businesses to begin developing and/or improving their FSMS so that they may also enroll in the STAMP program and gain customer confidence.

The benefit of the STAMP program that may be of the most value to the regulatory authority is the reduction in inspection frequency based on the demonstration of AMC. Annex 5 of the FDA Food Code, Conducting Risk-Based Inspections, provides an example of risk categories and assignment of inspection frequencies based on risk. The Annex suggests that regulatory jurisdictions may consider voluntary FSMS, such as AMC, to justify a decrease in inspection frequency for those operations. Jurisdictions are often faced with diminishing resources, limited staff, and increasing workloads. Because of these pressures, it is important to assess regulated facilities for practices that carry the most risk and require the most attention and to prioritize the inspection of those operations. Reducing the inspection frequency for facilities demonstrating AMC enables resource reallocation to riskier establishments, thus facilitating intervention and training where they are most needed to achieve long-term reduction in risk factors for foodborne illness.

Continuing evaluation of our program is important in improving incentives and award programs to ensure that programs are successful and effectively achieving the desired outcome. Indeed, our analysis of the individual employee AMC award discussed above—and identification of some of its shortcomings—is what led to the creation of the STAMP program. A pilot of the STAMP program led to many program and process improvements before the program was rolled out to our regulated food establishments. Both programs achieved the initial goal of recognizing food operators and operations that practice AMC. However, the over-arching goal of

AMC recognition is AMC promotion and expansion of AMC in the jurisdiction to increase food safety. Admittedly, the first round of AMC award recipients and STAMP enrollees were foodservice businesses that already demonstrated exceptional AMC without health department intervention or assistance. We were proud to have programs through which we could recognize and celebrate these accomplishments; however, it was important to promote the adoption of AMC to foodservice businesses that did not yet have FSMS in place.

To make a positive impact on food safety, the health department must promote AMC and should encourage food establishments to strive for AMC awards and programs. Our staff utilizes training materials, monitoring, and policy templates to assist facilities in incorporating AMC into their daily practices. When our regulatory staff can demonstrate that promotion efforts have resulted in new facilities demonstrating AMC, we can truly say that our programs have been a success.

### How to Establish Active Managerial Control Programs in Regulatory Departments

Just as restaurant owners are faced with competing priorities, health departments face similar challenges. Many health department staff conduct inspections across a multitude of programs, not merely foodservice operations, and are under constant pressure to balance assignments and prioritize tasks. This creates a challenge to the assessment and promotion of AMC, which may be viewed as a luxury service that may not be feasible given the need to complete other mandated services. Some barriers identified that challenge the regulatory authority assessment and promotion of AMC are:

- Time—Health department staff are constantly faced with increased workloads and diminishing resources.
- Resources—Staff may not have the tools and resources needed to communicate the concepts of AMC to foodservice operators.
- Understanding of AMC—Staff may not have a clear understanding of what defines AMC, how to assess AMC in a foodservice operation, or how to effectively communicate the concepts of AMC to foodservice operators.
- Priority—In light of competing priorities, AMC promotion may not be supported by program management or staff.

These barriers are real challenges that face health departments across the country. When conducting benchmarking for the STAMP program, we found only a few programs nationwide that have information, tools and resources, or promotion programs that are centered on AMC. The challenges to AMC promotion and assessment are most likely the reason that these programs or efforts are not more widespread and must be carefully considered and addressed when committing to AMC promotion activities. It is important to recognize that although these activities require the investment of time, resources, and prioritization, the benefits to the public health of the community at large make increasing AMC in a regulator's jurisdiction a worthwhile endeavor.

Effectively assessing and promoting AMC in a regulator's jurisdiction will invariably lead to an increase in AMC. As previously discussed, some benefits to the regulatory authority for increased AMC include an increase in long-term compliance, less enforcement action, less time spent at inspections, a potential reduction in risk categorization, and the potential for fewer complaints due to the absence of foodborne illness risk factors. It is important to share these benefits when seeking buy-in from internal stakeholders to counterbalance the challenges inherent with undertaking AMC promotion activities and to create a culture of AMC promotion among staff. In order to successfully foster a culture of AMC promotion, it is important to make the concept and assessment of AMC clear, make assessment and promotion of AMC easy, keep AMC promotion activities exciting by continuously improving and enhancing programs, and keep the culture of AMC thriving by making it a continuous topic of discussion for regulatory staff.

In order to begin creating a culture of AMC in our jurisdiction, we started small. We introduced the concept of AMC through a presentation of Annex 4 of the FDA Food Code, which describes the regulatory authority's responsibility to have an inspection program designed to assess AMC and to encourage and promote AMC within foodservice operations. This introduction helped solidify AMC assessment and promotion as a best practice supported by the FDA, the national authority on food safety. We engaged our staff to help define AMC in terms of real-life observations. Together, we defined what we believed to be the core components of AMC (policies, training, monitoring, and corrective actions). We reviewed each risk factor and intervention to discuss specific examples of AMC that would serve as control measures for each. These discussions led to the generation of an AMC guidance document to assist staff in their assessment of an operation's AMC during an inspection. The discussions and guidance document increased the regulatory staff's comfort level with having conversations related to AMC and making suggestions to enhance FSMS.

In order to aid regulatory staff in promoting AMC, it is important to keep things simple and standardized. A team of staff created an AMC Toolkit that could be easily utilized by all regulatory staff to introduce the topic of AMC to foodservice operators and provide templates for developing standard operating procedures and monitoring logs specific to their operation. Additional training materials, educational posters, and monitoring templates were generated for each risk factor so that staff could share tools and resources needed to demonstrate AMC in these areas. To increase access to these resources, all tools and educational documents were made available online; many are planned for translation into additional languages to make the information accessible to establishments with diverse language needs. Our staff is also generating an AMC training course, to be hosted by the health department, that will introduce the concept of AMC and share valuable techniques, tools, and resources for achieving AMC over the risk factors in an interactive classroom setting. We hope that by providing AMC information and guidance in a format that is fun, simple, and easy to employ, foodservice operators will have greater success in implementing AMC into their operation.

Another goal of our program is to easily measure the success of our AMC promotion efforts. To facilitate collection of this information, our inspection program has added a non-scorable assessment item related to AMC on each inspection report. This ensures that the inspector is prompted to assess an operation's AMC. These non-scorable items will provide a measurable determination of not only the level of AMC in our jurisdiction but also the effectiveness of AMC promotion activities.

Since AMC requires an extra investment of time and resources, it is important to keep staff excited and engaged in promotion efforts. Staff were involved at all phases with the development of our AMC assessment and promotional programs and materials. This served to utilize staff expertise to enhance the programs; it also challenged staff to find means through which they could communicate AMC information and encourage operators to adopt these practices. The creation of the AMC award program got our staff excited about discussing the concepts of AMC with operators and identifying examples of AMC during their inspections. Staff were eager to make these assessments so that they could present an operator with the AMC award and have the opportunity to provide positive feedback and recognize the operator's achievements. This award proved to be very meaningful to our operators who weren't accustomed to being recognized and celebrated for their successes during a local health inspection. Enrollment of a facility into the STAMP program was also a motivating factor for staff when assessing and promoting AMC. Enrollment in this program facilitates reduction in inspection frequency due to a demonstrated reduction in risk. It is exciting for staff to assist their operators through the enrollment process, see them excel, and know that the risk of foodborne illness within our community is reduced with the increase in AMC due to their promotion efforts. Additionally, staff that assist in the successful enrollment of a facility into the STAMP program are eligible for a health department award recognizing exceptional customer service, further incentivizing collaboration.

The most challenging part of any new initiative is generally keeping efforts sustained in the longer term. In order to keep AMC as an active consideration for our inspection program, we looked for ways to continuously discuss and assess our activities. AMC is a recurring agenda item for our monthly staff meetings, and we use this time to discuss successes and challenges for AMC assessment, promotion, recognition, or enrollment into the STAMP program. These conversations enable our staff to benefit from each other's experiences, share strategies that have been successful, and work together to develop solutions to any obstacles. These discussions also help inform the management team of changes needed at the programmatic level to make AMC promotion activities easier and more effective.

It is also important to keep buzz around AMC active with foodservice operators and the community in order to sustain interest in committing to AMC. We include information on AMC, our newest AMC awardees and STAMP enrollees, and helpful tips on how to incorporate AMC into foodservice operations in each quarterly newsletter sent to our foodservice establishments. We have also included information about AMC on our website, issued press releases related to our AMC programs, and engaged the community through social media. This ensures a jurisdiction-wide awareness of AMC and engages our citizens as partners in promoting and encouraging AMC. By keeping AMC as a part of our everyday discussions and continuously striving to improve our programs, our promotion efforts are thriving and sustainable.

**Summary**

Local regulatory inspection programs (health departments) are continually striving to be more successful in identifying risk factors and effective interventions and employing highly trained staff to make standardized observations and recommendations to foodservice businesses for enhancing food safety. However, in the absence of an inspection program designed to assess the level of AMC in a foodservice business, can we affirm that an inspection free of risk factor violations at the time of an inspection is successfully controlling these conditions every day? Compliance documented on a health department inspection report alone is not an indicator that proactive FSMS are in place to identify and control foodborne illness risk factors. Health department inspections, although effective at identifying risk factors at the time of inspection, are only a snapshot of how an operation performs every day. For this reason, it is important for the regulatory authority to incorporate the assessment of AMC demonstrated by a foodservice business into their inspection process. The regulatory authority must also encourage and promote the adoption of AMC practices using FSMS where they can be improved or initiated.

The FDA Food Code asserts that in order to effectively reduce risk factors, a foodservice operation must achieve AMC. Adopting AMC practices in a foodservice operation requires prioritization and an investment of time and resources. By the same token, assessment and promotion of AMC by the regulatory authority requires similar investments. With both industry and the regulatory authority constantly faced with competing priorities, it is important to recognize the value of AMC and commit to its prioritization with the common goal of producing safe, quality food for customers. The regulatory authority should utilize its expertise to help facilitate AMC in their jurisdiction. Trained inspectors may assess the degree of AMC in an operation and provide valuable feedback for improvement. The development of AMC tools, resources, and trainings will help simplify and encourage the incorporation of FSMS into an operation. Recognizing success by presenting operator awards or enrolling an operation into an elite recognition program helps to celebrate and incentivize the achievement of AMC.

Empowering and engaging staff in AMC activities will also be integral to the continued success of promotion efforts. It is important to ensure that regulatory staff are provided with adequate training and resources so that they are comfortable and eager to provide AMC assessment, feedback, and the tools needed to initiate FSMS. Keep regulatory staff excited about promotional efforts and keep the programs thriving by fully ingraining these conversations into the inspection program and creating a culture of AMC promotion. AMC should be recognized as a critical intervention against all foodborne illness risk factors. The health department shares responsibility in facilitating the incorporation of AMC into foodservice operations. By assisting operators in developing and implementing strategies to strengthen FSMS, the health department may serve as a partner with industry to play a significant role in the reduction of foodborne illness risk factors in the community.

**Shannon McKeon, REHS**
**Environmental Health Specialist III**
Activemanagerialcontrolpromo@gmail.com

# Index

© Springer Nature Switzerland AG 2020
H. King, *Food Safety Management Systems*, Food Microbiology and Food
Safety, https://doi.org/10.1007/978-3-030-44735-9